Business Skills for
General Practice
Surveyors

Austen Imber

2005

Routledge
Taylor & Francis Group

LONDON AND NEW YORK

First published 2005 by Estates Gazette

Published 2014 by Routledge
2 Park Square, Milton Park, Abingdon, Oxon OX14 4RN
711 Third Avenue, New York, NY 10017, USA

Routledge is an imprint of the Taylor & Francis Group, an informa business

ISBN 13: 978-0-7282-0432-4 (pbk)

Typeset in Palatino 10/12 by Amy Boyle, Rochester, Kent

Contents

Foreword

Surveying is a people person's profession where interpersonal skills and business acumen are essential. Successful surveyors are "all rounders", a quality that requires good people skills, confidence, presentation and financial acumen. Natural intellect is rarely a single factor. Lifelong learning commitment can, therefore, reap rewards, with the reading of articles and books such as *Business Skills for General Practice Surveyors* playing a key part.

Good business skills, in addition to technical skills, provides many things. Clients' and employers' combined business and property requirements can be accurately understood and investment and development opportunities can be exploited. Surveyors who continually develop their skills are also able to develop their own property investment interests alongside salaried employment, or establish their own property consultancy.

I have worked with Austen for many years and he enthusiastically works with young surveyors — providing motivation and encouragement to enable them to become more than just technically proficient. His no-nonsense, can-do style and attitude is a fundamental ingredient to inspiring the many people that he works with. He encourages surveyors to strive to achieve the best they can and to make positive changes to their day-to-day thought process and long-term career prospects.

Austen asked me to write this foreword because I became a partner of an international property consultancy at the age of 28, and have both client-side and private practice experience. In addition, a career in property has enabled me to develop a personal investment portfolio in the UK and overseas. I strive to have a good business instinct and strategic grasp, and, as a result, I am always pleased to promote personal development and lifelong learning.

My message to all young surveyors is to master their property skills through university and the RICS Assessment of Professional Competence, and to make sure that their careers continually develop thereafter. I hope that this book helps surveyors to be bold and confident, and to make decisions that they have long pondered.

Austen really does practice what he preaches. His career has an unusual combination of property investment, surveying, business advisory and training interests — affording a "lifestyle" existence from his early 30s.

Mrs Kerry Bourne
Partner, GVA Grimley

Preface

Technical knowledge is only the starting point to being a successful surveyor. Property investment and development is about making money, and the corporate sector requires property strategies to fit the operational nature of their individual businesses. All of the profession's clients demand high performance from their property advisers. Business has to be won in private practice, and the public sector has to meet performance criteria with property dealings.

Within the property sector, there are successful businesses and unsuccessful businesses. There are high-earning surveyors and lowly paid surveyors. Business acumen, entrepreneurial spirit, attitude to skills development as well as an appetite for hard work are common themes accounting for such contrasts. Aside from the few high-flyers and those with fortunate connections, younger surveyors are on a broadly equal footing when looking to accelerate their own performance and earnings capacity. Furthermore, for those surveyors with strong business acumen and a good understanding of particular sectors, industries and businesses, property advice can be complemented with business advice. Experienced surveyors will be at various stages of their career but, for example, a managerial position in mid-career can provide a focus and kick-start to greater things. Indeed, whatever the current pay of individual surveyors, or earnings and efficiencies within a company, effectively embracing business skills pushes careers and businesses forward.

Business Skills for General Practice Surveyors provides a grounding in business issues relevant to surveyors' case work. Comment on surveying matters is kept to a minimum.

Chapter 1 explains the application of business and management skills for general practice surveyors in their various areas of work. This includes RICS initiatives to enhance business skills within the profession, and points highlighted in the government's Egan skills review. Initial illustrations are provided in respect of clients' and employers' demand for business skills.

Chapter 2 provides an overview of the property market, explaining the structure of the industry, and the roles of its many participants. Development and investment processes are considered, including the opportunities and threats presented in changing market conditions.

Chapter 3 illustrates the business dimension to surveyors' general practice case work, and concludes with a summary of business skills derived from the various examples included in the chapter. Chapter 4 explains the management consultancy roles undertaken by surveyors and, again, includes various case examples.

Chapter 5 explains the workings of the economy, the economic statistics often reported by the press, and how the UK property market is influenced by economic conditions. Illustrations are provided of property market trends and market information on which surveyors can draw. Chapter 6 provides an overview of the equity market. Chapter 7 helps surveyors understand company accounts and other financial information, and how this affects companies' current trading performance and future opportunities.

Chapter 8 provides guidance on negotiating techniques used in day-to-day property dealings, and which are essential in getting the right result in case work. Chapter 9 covers presentation techniques, ranging from large audiences to small groups.

Chapter 10 initially comments on how training activity adds to the necessary business skills, as well as keeping up to date with changes in market practice, legislation and such like. It then provides a reminder of the primary sources of success for individuals and businesses — entrepreneurial creativity, the ability to read markets and exploiting gaps in markets. The chapter also briefly summarises the factors that contribute to the successful internal running of property-related businesses, including the concept of "total business management". However, detailed guidance on the running of businesses is not provided.

The above subjects provide an important reminder of the harder surveying and business skills that contribute to business and personal success. This is particularly important when general business training opportunities for surveyors are often devoid of a property context — with consultants and hired trainers instead drawing on more abstract content and methods of training in order to help widen their markets, and overcome a lack of knowledge of a client's industry and related business issues.

By providing a grounding in business issues affecting general practice surveyors' case work, this book will help surveyors wishing to undertake future learning in individual subjects. This includes softer skills, which are confined in this book to primary subjects of negotiation skills and presentation skills, which have a direct impact on surveying case work.

Although the term "client" is commonly used in this book, this also means employer in an in-house sense — as in the public and corporate sectors, there may be internal, departmental or operational clients to whom the surveyor reports.

For surveyors considering setting up their own surveying practice, a separate Estates Gazette Books' publication is available: *Starting and Developing a Surveying Business: For Sole Traders*. Many of its chapters will also benefit existing businesses wishing to examine issues affecting their future development and profitability. *Starting and Developing a Surveying Business* also outlines the pros and cons of trading status, such as sole trader, partnership, and limited company, and taxation issues of income tax, national insurance, corporation tax and VAT — all of which are outside the scope of *Business Skills for General Practice Surveyors*.

Thanks are due to RICS for the use of case examples, especially RICS Chief Economist Milan Khatri in respect of chapter 5. Case examples and other extracts from RICS Management Consultancy faculty are included in chapter 4, and thanks go to Margaret Harris of RICS, not only in respect of extracts in chapter 10, but also her work on behalf of RICS in promoting business skills. In addition, the work of Scott Kind as training and development manager for GVA Grimley is acknowledged, and which is featured in chapter 10.

Further thanks are due to Alison Richards, EG Books' commissioning editor and to Rebecca Chakraborty, marketing manager of EG Books. Essential support in respect of proofreading, production and typesetting has been provided by Howard Imber, Isabel Micallef and Amy Boyle respectively.

Austen Imber

Introduction to Business Skills

In the early stages of a surveyor's career, the development of technical knowledge and the ability to apply this to case work tends to be the focus of learning and personal development.

As a career progresses, business and managerial qualities become increasingly important in gaining promotion and pay increases, and in being able to find the preferred areas of surveying practice.

This combination of property and business expertise reaps rewards for individuals because it helps facilitate optimum performance from a property-related business. Examples include a property consultancy serving its clients, a public sector property team delivering its strategic objectives, a corporate occupier ensuring that property decisions accurately reflect operational requirements, the corporate and public sectors securing property income and capital for their operational arms and property investors and developers making as much profit as possible.

RICS and business skills

RICS Agenda for Change in the late-1990s reiterated the weakness of business and managerial skills within the surveying profession, and the need for improvement. This was considered necessary in order to maintain and enhance the standing of chartered surveyors within their client markets and, importantly, understand clients' particular business objectives and requirements, thus enabling quality property advice to be provided.

The Agenda for Change was also conscious of surveyors' wish to command higher fees for their businesses, and increased salaries for themselves. Other aspects included the need for the surveying profession to be defensive against the scope for other professions to take over areas of work traditionally undertaken by chartered surveyors. RICS also aims to be the leading worldwide body regarding the provision of property services — due in part to the benefits considered afforded to members and individual countries from a global profile.

The "profession's clients" is a broader concept than simply the clients of property consultancies, and incorporates RICS corporate and public roles. RICS and its members' work in the property industry contributes to the success of the UK economy, government policy making, planning and environmental issues, legislative reform and consumer protection.

Property issues have recently received more attention by government, the media and the public than has been the case traditionally. This reflects the increased value of property assets, and also factors such as the impact of house prices on the economy, the contribution made by property to regeneration, the benefit to commercial occupiers of a flexible leasing market, and the increased use of property as part personal investment and financial planning. RICS' promotion of chartered surveyors as part of the Agenda for Change also plays a part, as do the number of television programmes on property (albeit concerning mainly residential property).

In terms of surveyors' education, RICS has looked at introducing the requirement for surveyors to take a formal business qualification (termed a "post-graduate business qualification") or encouraging learning in business subjects in other ways — such as surveyors' Continuing Professional Development (CPD). This builds on RICS implementation of the "lifelong learning" theme, which means that at all stages of their careers, surveyors should plan education and training activities that will be beneficial from both a business and more personal career perspective. RICS stresses the value of business skills being developed once surveying experience has been gained in practice, rather than business skills featuring too heavily on degree courses and displacing the core property areas. Whether or not a business qualification becomes mandatory, RICS education and training department will continue to promote business skills through a series of initiatives.

RICS' promotion of the surveying profession, together with its cyclical tendency to attract school leavers when the property market is performing strongly (and there appears to be money to be made) has recently attracted increasing numbers to consider a career in property. In contrast with the profession's difficult times in the early to mid-1990s, when some universities relaxed student entry and other standards in order to survive, the quality of graduates entering the profession has generally improved. This includes a rise in the number of non-cognate entrants, including some with superior credentials to the mainstream intake, and many with a business-related degree. The scope to gain a part-time property degree and reach chartered surveyor/MRICS status within two years (as the degree runs currently with the Assessment of Professional Competence) is an obvious attraction for non-cognate degree holders.

RICS Management Consultancy faculty takes responsibility for others areas of business skills development in view of its members' areas of property work being specifically related to the operational nature of clients' businesses. The work of RICS Management Consultancy faculty is included in chapter 4.

Government skills review

In 2003, the government commissioned the built environment skills review, otherwise known as the "Egan skills review".

Its purpose was to review the skills needed to deliver regeneration and develop sustainable communities, and, in particular, find ways of enhancing skills where deficiencies were found. The review covered over 100 occupations, of which only property is part, but nevertheless highlighted areas within the property sector that contributed substantially to regeneration, and therefore the success of the wider economy and the general pursuit of economic growth.

Key areas identified for skills development were business and management-related, including strategic planning, leadership skills, financial awareness, people management and communication skills. Although the many other factors that are relevant in delivering regeneration relate to the performance of the government itself, to the attractiveness of working in the public sector, planning and compulsory legislation, state aid/funding rules etc, the focus on business skills is still consistent with RICS initiatives regarding business skills. Ironically, in the regeneration sector, and as a further example of the rewards from developing business skills, people in public sector regeneration bodies who demonstrate strong business and managerial acumen can often be poached, and paid more, by the private sector.

The need for certain skills in the regeneration sector illustrates the importance of business and management acumen in public sector property teams as well as in private property consultancies. This includes the Regional Development Agencies (such as Advantage West Midlands), English Partnerships (the national regeneration agency), local authorities and many other organisations.

Another aspect of surveyors' work reinforced by the Egan skills review, and summarised by RICS representation to the review, is the common need for surveyors to understand their particular field of expertise in detail, but also have a base understanding of many other property, economic and people issues — including the roles of individuals and organisations and how they contribute to delivering objectives. This was in the context of regeneration, where strategic vision and teamworking are key elements, but it characterises many other areas of the property sector.

Understanding the property industry and its clients' needs

As surveyors tend to work in a particular sector, or are involved in certain areas of work, their knowledge of how the whole of the property industry works can sometimes be limited.

It is important that surveyors understand the dynamics of the property market and its relationship with the economy. The outlook for the economy, the scope for interest rates changes, trends in localised submarkets, etc, influence investment and development risk, and the advice given to clients or decisions taken on behalf of employers.

When undertaking a letting on behalf of a landlord investor, for example, the specialist agent must be able to evaluate the client's investment objectives, such as an imminent sale to take profits, or the wish to refinance and extract funds to expand the portfolio by buying other properties. In both cases, capital value will be key, with specific regard being given to tenant selection, a sufficiently long lease term and other lease provisions that are investment/finance-friendly. Similarly, a landlord and tenant specialist instructed to undertake a rent review in a multi-let property, or where a landlord has a portfolio of interests in an area, needs to be conscious of the effect of the rent review, as comparable evidence, on the rental levels achievable on other rent reviews and lease renewals in the building/ portfolio. It may also be beneficial to delay the rent review pending the emergence of better comparable evidence (as opposed to looking to conclude the instruction as soon as possible in order that fees can be billed).

A surveyor who is able to reflect the client's objectives within the surveying services provided, and achieve successful results, will be valued more highly than a surveyor appearing to provide generic advice through standard processes. As a result, more work is likely to follow and higher fees should be commanded.

Private clients are usually in the business of making money, and want their advisors to embrace the same culture. The public sector may be particularly conscious of best consideration/best value, transparency in property dealings and potential political difficulties. The corporate sector may be sensitive to public relations disasters that can affect their operational trading performance (ie "reputational risks"). Such examples further illustrate the business skill of understanding the client/employer and the nature of their business.

Surveyors tend to consider property as the primary element within their work, but businesses see property as secondary to the factors that drive their sales, profitability etc. Non-property issues may override property issues, despite the apparent value of property and the surveyor's view that optimum use is not being made of property, and surplus assets are not being exploited. As a simple example of the business dimension related to property, a city centre office occupier with staff costs (salary etc) averaging £30,000 per employee, working to an average of 100 sq ft space per person, has a staff cost of £300 per sq ft. This puts into context the relative pre-occupation by surveyors of a few pounds and pence difference in occupational costs. This accounts for why many city centre office occupiers tend to gravitate to the best space — optimising profile to clients, and the working environment for employees at relatively small additional cost. In other situations however, property costs are more significant to the occupier, not least because of their sensitivity to property margins, if not total cost, and because property is not being managed efficiently. Such issues are covered in more detail in chapters 3 and 4.

Own operational performance

In addition to the client-facing elements of business skills, surveyors working for property consultancies (especially those with managerial responsibility) will need to understand the economics of their business in terms of revenue/fee income levels, fixed costs, variable costs, profit margins, the scope for cost savings, the pricing of services and the risks and returns associated with investment expenditure. Business plans may have to be prepared, fee targets set and a certain level of profitability achieved by the year-end.

The same principles apply to other businesses, such as a local authority property team where resources may need to diminish proportionately to the reduction in the council's property portfolio due to sales — as opposed to the same number of people being retained, and the organisation and individuals being unaccountable in terms of productivity, case lists, performance measures etc. Surveyors also need to be alert to the risk of outsourcing, and a private firm, for example, stepping in and immediately identifying the scope for job cuts.

Businesses need to keep up to date with developments in information technology. As well as maintaining efficiencies, this can help exploit new business opportunities and give some property consultancies and surveyors an edge in their markets.

Professional advisers will usually handle such matters as employment law, discrimination legislation, directors' duties and various other statutory requirements, but surveyors need to be aware of the key aspects and how they affect the running of the business.

Excelling in property and business

As mentioned in the preface, it is the combination of property skills and business acumen that drives surveyors' careers forward. Many successful people are not the most technically qualified, and the extract below from the section of "Successful business people" in the separate Estates Gazette Books' publication, *Starting and Developing a Surveying Business*, summarises the key points.

As with all business people, the surveyors who generally do best are those with an entrepreneurial eye. They instinctively see gaps in the market, and understand how sub-markets and niche concepts can be exploited. A combination of technical knowledge, expertise in practice, an ability to read markets and basic business skills creates an edge on others. Incisive thought processes and the ability to carve out sharply focused concepts see strong demand for the services on offer — the profile and exclusivity of which cannot be matched elsewhere. Intelligence is not the same as thinking deeply, analytically, creatively, and with vision — nor being able to think quickly and correctly, especially when under pressure.

The instincts of highly successful business people are hard to explain, and difficult for others to understand. They just "seem to know", and their strategies reap success. While the masses follow the herd, successful business people focus on ways to add value to existing services/ventures, or create fresh themes. Instinctively, successful business people see how relationships between revenue, costs, margins, etc, combine.

Psychology is important in business: such as how to captivate consumers/clients, how different clients make decisions to appoint advisers, what motivates staff and how the market perceives a business/brand. Confidence is also a key element in driving forward a business, just as it is in the performance of economic and property markets. The building of relationships, use of contacts and ensuring trust among all those with whom business is conducted is important.

Surveyors should still draw on the traits of highly successful business people, and apply elements to their own business. As indicated above (in *Starting and Developing a Surveying Business*), high financial rewards are achievable, but that does not belie the difficulty in formulating the right concept, and being able to win sufficient business at the right price.

To a large extent, the laws of supply and demand determine the viability of setting up in business. In certain locations, and with certain areas of existing expertise, the venture will simply be unviable. However, one surveyor does better entering an already crowded market because of particular initiatives deployed, than another surveyor with niche services of great potential, but not the necessary skills to win business or manage the venture effectively. Similarly, the assessment of the interrelationship between certain services, fees, working methods, costs, margins and profits (for both the business as a whole, and individual fee lines) sees highly profitable ventures established, where others cannot see such dynamics.

Further illustrations are shown in the section "Becoming brilliant" in chapter 10. The final section in chapter 3, p42, summarises the skills typically required by surveyors.

Property Industry Overview

It is important for general practice surveyors to understand the dynamics of the property market and how their individual areas of practice fit in.

The performance of the property market is linked closely to the performance of the economy. Inflationary pressures, for example, will be met by higher interest rates, thus increasing not only finance costs for property investment and development, but also increasing the yields required by investors, and therefore putting downward pressure on capital values. Decisions to acquire investment property, commence development or expand into new premises carry risks that the surveyor must understand, and these need to be conveyed articulately to clients.

Capital and rental valuations of property are influenced by activity in the investment, development and occupational markets (retail, office, industrial, leisure etc). Investment and development performance is reliant on the strength of occupational markets, which in turn are influenced by factors such as consumer spending, the expansion of the service sector, and the competitiveness of manufacturers in view of the strength of the pound.

Oversupply and lack of supply can be localised and small submarkets can perform very differently from the national trends. The performance of the financial markets influences property in many ways, including the effect of city bonuses on parts of the London residential market, asset allocation (including property) by fund managers and general market sentiment/confidence.

An up-to-date knowledge of current issues is required, and how property securitisation, for example, could influence investment and development markets — ie Real Estate Investment Trusts (REITS)/Property Investment Funds (PIFS). Although governments introduce legislation which provides opportunities that can be exploited, it may also bring risk and financial losses to existing areas of business.

Surveyors need to understand the roles of other surveyors, other property professionals and anyone else with an interest in the property market. This enables multi-disciplinary, team-based project work to progress effectively and, in the case of a property consultancy, helps facilitate the cross-selling of services by drawing on colleagues.

Surveyors need to appear knowledgeable to their clients and colleagues, demonstrating an understanding of market trends and all areas of the market. A commercial property surveyor's detailed understanding of the residential market, despite not undertaking such work, may be particularly valuable in a personal sense to a client interested in acquiring private residential investments. Incisive residential market and economic commentary can also flatter non-property clients as to the surveyor's general acumen, and most people have an interest and base understanding of the residential market.

This chapter provides only a basic overview of the property industry. More detailed illustrations of the issues facing the types of client/organisation surveyors may act or work for are included in chapters 3 and 4.

Development land and viability of development

In simple terms, the property process begins with land.

The land may be "greenfield" land, such as agricultural land on the edge of town, or school playing fields, which have not been built on previously. "Brownfield land" has already been developed previously, such as a factory and distribution yard in a town centre that has recently closed. Therefore, "land" really means "land and buildings", or a "development site".

In order for any development to take place, planning permission is needed. The land could be owned by an individual or a business that wishes to sell for the best price, or could be owned by a property developer who acquired the site previously. In order for development to be commercially viable, the value of the completed development will need to exceed the cost of buying the land and the cost of constructing the development — and also be worthwhile for the developer in terms of producing the required level of profit. If, for example, a developer can acquire an office development site for £1m, build the offices for £7m, sell the completed offices for £10m and requires £1m profit, then development will be considered commercially viable. However, risks include costs eventually being more than £7m and market values falling from £10m over the construction period — in both cases eroding profit and possibly leading to losses.

Sometimes, development may take place that is not commercially viable. A regeneration agency may fund the gap between end value and development costs (including developer's profit) in order to achieve development that supports the economic regeneration of an area. This will usually be for a site within a regeneration scheme, and although early developments may warrant grant funding, later developments will not need funding, as improvements to the area increase values and make development commercially viable. The regeneration agency can then recoup initial investment expenditure through land sales at higher, positive values. A business may also construct property in order to expand, despite this not being viable from a commercial property perspective, but because the profitability of the related business is so substantial.

Whether development is feasible also depends on the existing use of the land and its value. The existing use value of greenfield land will usually be minimal in comparison with alternative use value, such as £2000 per acre for agricultural use compared with £1m per acre residential use. A ground level, gravel surfaced, city centre car park may have an existing use value of £2m, reflecting the level of revenue generated, but if the alternative use value for an office development is £1m, such development will not take place.

Changing markets: opportunities and threats

It is important for surveyors to be alert to market trends and to generally be able to "read" markets. In the aforementioned example, the car park may be on the fringe of the city centre. However, the increased popularity of "city living" (town and city centre residential development) and the related development of retail and leisure uses may later see the site worth £3m. As such, alternative use value exceeds existing use value, and when planning permission is secured, the car park closes, and the site is sold. This loss of car parking space will have other effects on the area. Greater demand for other car parking spaces will mean a rise in daily car parking charges, an increase in total revenue and also site values. Shoppers who are now struggling to park when visiting the city centre and/or who are discontented with the increased cost of car parking may switch to public transport — or may decide to shop elsewhere. Retailing may therefore suffer to some extent, and this will be reflected in property values, investment returns and development viability.

The increased viability of residential use, reflecting low interest rates and rapidly rising house prices, is likely to see more secondary office space converted into flats. This will squeeze the supply of office space and is likely to increase its market rental value — beneficial for investors, but not occupiers.

Increased residential space in the city centre should have a beneficial effect on retailing and leisure markets. However, the prospect of gains for retail and leisure operators can see more development take place than can be supported by the influx of new residents, thus diluting the returns to individual operators/outlets and therefore also affecting property values. It is important to evaluate the potential chain reactions in local economies and property markets — however difficult this may be to do with accuracy.

A property developer can find sites and start construction, sell and take profits, and be out of the market by the time it falls. In all markets, valid decisions by individual participants can have detrimental collective effects, and account for boom and then bust. The oversupply of city living residential use, and of new retailing and leisure uses, are good examples. Confidence and sentiment can drive markets forward in good times, but also can hold them back in difficult trading conditions. Those who can anticipate market trends and read markets can make considerable financial returns out of property — with the key being the early anticipation of market movements, rather than market entry as activity begins to slow. As with the financial markets, there are always investors who pick their moment to enter a market too late in its cycle. The same principles apply, to some extent, to the market for surveyors' professional consultancy services.

Commercial development — including investment and lettings

The process for commercial development will usually be: a developer acquires a site, constructs the building, finds tenants and then sells to an investor (with planning permission also being in place). As well as keeping to the projected build costs and not incurring delays in construction, profitability will be influenced by the usual factors affecting the value of investment property — its level of income/rent, security of income/rent, scope for income/rental growth and scope for capital growth. The covenant strength/ financial standing of tenants, lease length, absence of break clauses, the precise nature of certain other lease terms, whether single let or multi-let, and other factors will be important. In other situations, the developer will hold the property rather than sell. This could be because their role is one of developer/investor, or because it is better to wait until the property is fully let and full market rents are established before selling to the investment market and thus optimising profitability.

Rather than build speculatively, the developer may have secured a pre-let, such as an anchor tenant supermarket for part of a retail scheme, or pre-lets for all or most of the building. Although this may mean that the opportunity is denied to obtain the best possible rents at the time that the development is nearing completion, it will reduce risk and be beneficial from a financing perspective (bearing in mind that most developers are reliant on finance). Rents, for example, can be fixed, and rental voids minimised (although there may be some rent-free period incentives). Pre-lets to tenants of particularly good covenant strength reduce the investment yield and increase the capital/investment value. Also, the reduced overall risk enables a developer to accept less profit. Such factors associated with pre-lets (including variations to a number of inputs within a development appraisal) help a developer to pay more for the land, and therefore be the successful/highest bidder.

Commercial property, the economy and planning

As detailed above, the amount, type and timing of commercial development depends on factors such as the ease of winning planning consent and the strength of the property investment market.

Property serves many economic functions. This includes its occupational function in the running of businesses (and the contribution businesses make to economic growth). The cost of property, as with all business costs, affects the international competitiveness of UK businesses and the economy as a whole. Property's investment function is alongside other asset classes, including shares/equities, bonds/gilts and cash/deposit accounts. This is not only at institutional level, but also increasingly as part of personal financial planning at consumer level.

The planning system is often criticised as being slow and inefficient, and that this hinders new development. Although often bemoaned by property developers, in most markets where supply is constrained, it is a characteristic that enables high profits to be made. In contrast, in markets where supply has the scope to easily expand, prices and profitability can fall. In the property sector, supply could expand quickly, for example, because of relaxed planning controls on city centre living/residential development and conversions.

The planning system and the pace of development also has to reflect environmental issues, sustainability and such like, and the need for due process in terms of consultation and public interest. As a result, legislation needs to balance such factors with the economic need for development.

Housing and the economy

The planning system also influences the number of new homes that can be built, the easing of the housing shortage, and, in again adding to supply and reducing prices, the affordability of homes, particularly to first-time buyers.

As outlined in chapter 5, interest rates are set at an appropriate level in order to best manage the economy. In particular, inflationary pressures need to be curbed by higher interest rates. The many effects of higher interest rates include reduced consumer spending power (as more income is incurred on interest repayments), downward pressure on house prices and therefore less scope to borrow against house prices, increased cost of development and investment finance, downward effect on property investment and development values, increased costs of borrowing for businesses, and an increased strength of the pound (which makes exports more expensive and less competitive, causing manufacturing output to fall, jobs to be lost and consumer spending to decline). This is just a small illustration of the interrelated elements and circular effects taking place within the economy and the property market. It also highlights the precision with which governments need to manage their economies, with international markets further influencing the outlook for a domestic economy. Further detail is provided in chapter 5.

The surveyor's eye for enhancing value

Alternative uses should always be in surveyors' minds when dealing with property, as should all opportunities to get the best out of property. In addition to the development uses already mentioned, a change from general retail/A1 use to café-takeaway-restaurant/A3 use could increase the market rental value of a shop substantially, as well as helping add to the "footfall" (people passing by) for the owner's neighbouring shops. Surveyors also need to search out opportunities for additional and short-term income, such as a telecoms installation on the roof, an advertising hoarding on the side of the building or renting out car parking spaces.

Property consultancies

The various terms used in respect of firms who provide property advice to clients in a "private practice" capacity include property consultancies, international real estate advisers, chartered surveyors and property advisers.

Some of the large property consultancies are public limited companies (plc) quoted on the UK stock market, with their shares being available to the public (see chapter 6). There are UK firms owned by larger American-based companies. UK surveying practices may also be part of larger UK firms. Some of the main UK firms are partnerships, whereby the principals and staff share in all the profits of the company, as opposed to profits being distributed to parent companies or shareholders. Some firms are limited liability partnerships.

Acquisition and merger activity takes place among the large practices. If a firm is quoted on the stock market, it could be subject to a takeover, unless there is a majority shareholding/majority of shareholders resisting this. A partnership is private and would be bought out only at its own wishes. Mergers sometimes take place between the larger firms, and as well as expanding through running their business successfully, the larger firms may seek growth through the acquisition of smaller firms.

As an example of the scale of the larger practices, published statistics in June 2004 showed that FPD Savills has a UK turnover of approximately £200m, CB Richard Ellis £150m, Jones Lang LaSalle £120m, DTZ Debenham Tie Leung £120m, Knight Frank £95m, King Sturge £85m, GVA Grimley £80m, Lambert Smith Hampton £70m, Colliers CRE £60m and Chesterton £55m. Profitability measures are complex as this depends on how directors'/partners' salaries, bonuses, profit shares are treated (as well as other accounting issues). However, as a broad indication, profit margins tend to be in the region of 10–15% (eg a £100m turnover and £15m profit = 15% profit margin). The smaller firms, with a turnover of between £1m and £10m tend to have more scope to secure higher margins, and this may extend to 25% and sometimes higher for niche operators. The statistics on global turnover included CB Richard Ellis at £1130m and Jones Lang LaSalle at £580m. Also, "international" does not necessarily infer a large-scale, worldwide operation, as some of the large UK practices secure relatively nominal presence in overseas markets in order to support their UK profile.

Within the towns and cities throughout the UK, there are many smaller firms comprising a handful of surveyors, and earning similar profits year-on-year without any particular aspirations to expand. There are also small firms looking to expand aggressively through the efforts of particularly committed principals. Self-employment and working as a "sole trader" is the preferred combination of lifestyle and working methods for some surveyors, and there are also numerous two-person partnerships.

Surveyors also work with firms who are not regarded as property consultancies, but who are in the same market and serving similar clients. Accountancy practices, for example, have property teams, and firms involved in management consultancy within the property sector (see chapter 4) may profile themselves more as business consultants than surveyors.

The corporate sector

The corporate sector broadly means non-property businesses, but more specifically, larger companies and organisations.

Examples include retailers with outlets throughout the country, multinational manufacturers occupying industrial premises, banks with offices and high street outlets throughout the country, utilities with nationwide infrastructure, and the railway industry with its portfolio of railway station retailing and old goods yards suitable for redevelopment.

All businesses with property holdings need to make the most of their property assets, but with the larger concerns, there is more of a need to manage property strategically. For smaller businesses, property is more likely to be handled on a property-by-property basis, often with only one property being occupied.

The corporate sector will sometimes have an in-house property team that handles the usual range of general practice surveying requirements. Property consultancies will still often be appointed for some activities, and some corporates will rely heavily on property consultancies and have only a skeleton property team if, indeed, any chartered surveyors. As indicated in chapter 4, the corporate sector does not always place property highly on its agenda.

The public sector

Parts of the public sector are similar to the corporate sector in the sense that they may have large operational property portfolios, but also they have property teams who manage property that is surplus to operational requirements.

Local authorities and county councils employ property teams, although some outsource the property work almost entirely to property consultancies. Regeneration agencies have their own surveyors, while also working closely with property consultancies for certain requirements. Various government departments handle property responsibilities through in-house surveyors and appointed property consultancies. The education sector and the National Health Service are good examples of where property has been reviewed in order to establish the potential for disposal receipts and the effective re-use of funds.

Property investors, developers and construction companies

Property investors range from institutional investors, such as pension funds and insurance companies holding portfolios worth billions, to individuals buying a second home to rent out as part of their personal financial and/or retirement planning.

Institutional investors will usually hold property alongside other asset classes, such as equities/shares, bonds/gilts and cash, and have a low-risk profile. In the case of property, the better-quality properties will be sought, such as a £50m city centre office investment providing a 6% yield.

Property companies include major investors quoted on the stock market, private investment companies and smaller traders. As well as buying some lower-risk investments, they will be looking for more pro-active opportunities to make money. Secondary offices requiring some refurbishment, the letting of voids and the completion of outstanding rent reviews and lease renewals, for example, will provide opportunities to secure increases in both rents and capital values.

Private investors are also a significant part of the investment market, including high-wealth individuals. Overseas investors include institutions, property companies and individuals, all of which have a significant aggregate effect on market trends, pricing and such like.

Some developers will simply find sites, develop and sell, as mentioned earlier in the chapter, but some institutions and larger property companies will combine development and investment interests. Construction companies will generally make their money from building properties on behalf of developers (and also owner occupiers), but some may have their own development interests — and, similarly, developers may have their own construction team and not employ external contractors other than for specialist services.

Within the development processes, general practice surveyors (including planning and development surveyors) will handle development appraisals, land acquisitions, land sales, planning applications, lettings, investment sales etc. Once tenants of an investment property are in place, landlord and tenant work (rent reviews, lease renewals) and estate management (rent collection, alienation, dilapidations and other lease breaches, service charges, insurance, health and safety etc) will be a key driver of investment performance.

Lenders

Property investors other than institutional investors will usually look to raise finance to fund their acquisitions and also to make more money. If, for example, an investor has £1m available, a £1m property could be purchased without finance or, alternatively, £1m could be the investor's equity within a £4m acquisition, with a lender, such as a bank, providing £3m. This is a "loan to value ratio" of 75% (£3m debt/finance ÷ £4m property value).

In simple terms, if properties each yield 8% rental income and benefit from an increase in capital value of 10% in the first year, the cash-only (£1m) investor makes a total return of 18% (£80,000 income, £100,000 capital gain — ie £180,000 ÷ £1m).

If finance is available at 6%, the financed (£4m) investor makes a total return of 54% on his capital outlay/investment of £1m. This is a rental return of £140,000, calculated as 8% of £1m equity (£80,000), plus 2% (8% rent less 6% interest) of £3m debt (£60,000) and a capital return of 10% on £4m (£400,000). The higher the level of borrowing or "gearing", the higher the potential return if things go well, but the greater the risk of losses if things go badly (such as rising interest rates/finance costs, economic downturn, rental voids, lower rents). Also, investors can draw on increases in capital value (through refinancing) to acquire more properties.

Acquisition and management costs would also have to be considered in the above example, but there would also be scope for increases in rental value over time.

The bank and other sources of lending have an important effect on the property market, with the suitability of a property for loan security/lending influencing its investment demand and capital value. In the residential market, property values can rise sharply when interest rates fall, with affordability to owners, and the prices they are prepared to pay, relating more to monthly repayment costs than to the actual capital value/purchase price of the property. Lending is also important in property development, funding the cost of site acquisition and construction. Chapter 3 provides further illustrations of the business issues that property investors and developers face, and therefore concern surveyors acting on their behalf.

RICS

The Royal Institution of Chartered Surveyors (RICS) has around 110,000 members, which increases following mergers with other property-related bodies, an expanding UK property sector and RICS worldwide membership development initiatives.

As part of the Agenda for Change, RICS is positioning itself as the leading worldwide body in respect of property, benefiting its UK members through the global profile created, and the opening of overseas property markets to formally recognised, RICS-related, property consultancy roles.

General practice surveying is only one area of practice represented by RICS, although this tends to be the most dominant, accounting for nearly 50% of total membership. RICS represents the differing areas of practice through its 16 faculties, which are: antiques and fine art; building surveying; commercial property; construction (quantity surveying); dispute resolution; environment; facilities management; geomatics; management consultancy; minerals and waste management; planning and development; plant and machinery; project management; residential property; rural and valuation. These are not specific to types of surveyor, such as a general practice or planning and development surveyor, and an individual surveyor's work often covers a number of faculties.

The other principal types of surveyor are quantity surveyors (around 35% of RICS membership and building surveyors (approximately 10% of RICS membership).

There are other forums and groups, such as the telecoms forum, regeneration forum and the corporate real estate group. The Investment Property Forum established by RICS has developed a good profile within the investment community, including research and education programmes for its membership of surveyors and other professionals.

As a further indication of the profile of RICS membership, around 50% of members work within property consultancies/private practice (and have clients), 15% in the public sector, 15% within commercial concerns (corporate sector, investors, developers) and 15% are retired/non-practising. There are other small categories, such as education and research, and there are also chartered surveyors who work for RICS.

RICS works with around 40 "partnership universities" in the UK, which provide full-time and part-time degree courses. Together with the Assessment of Professional Competence (APC), this results in "chartered surveyor" status and the designation MRICS (member of RICS). The APC comprises a minimum two years' work experience/professional training, ending in a final assessment process comprising written submissions and an interview.

Other professionals

General practice and planning and development surveyors work particularly closely with solicitors in respect of property transactions (sale contracts, leases, development agreements), litigation (recovering rent arrears, dilapidations claims, other breaches of lease terms, lease renewals etc) as well as many other property and business areas.

In local authorities, planners will be members of the Royal Town Planning Institute (RTPI). Property consultancies and other employers, such as development companies, also employ qualified planners, some of whom hold dual RTPI/RICS qualification.

Accountants' support is often required in respect of property-related taxation and accounting issues, and regarding general business issues.

Business Issues in Property Dealings

In almost all general practice surveyors' property dealings, there is a business dimension. This chapter illustrates key issues that surveyors need to consider within their case work in order to provide accurate, good quality, property advice, which reflects clients' business operations and objectives.

The chapter begins with an example of the issues considered by the surveyor on behalf of a tenant looking to relinquish space, and shows how a range of general practice surveying functions combine with business issues. Further illustrations are then set out under sections of agency, property management, valuation, local authority property management, regeneration, public relations and marketing, property development and property investment, telecoms and consultancy roles. A case example is given of the combined business and property advice provided to a family property company looking to expand. The chapter concludes with a summary of the business skills which are highlighted from the various illustrations, and which are further considered in subsequent chapters.

Chapter 4 considers what is meant by "management consultancy" in the property sector. In simple terms, this is where surveyors advise businesses on the strategic management and utilisation of their property assets having specific regard to the operational nature of their business. A number of other terms are also used in the market, such as "corporate estate management" and "strategic property consultancy".

The "strategic" element is important, as management consultancy considers a business' property issues in an overall sense and determines a suitable strategy. In contrast, general practice surveyors' work is usually in connection with individual property transactions, notwithstanding their need to still reflect clients' business issues.

Although management consultancy work tends to be undertaken on behalf of larger companies with more substantial property holdings, there is not really a clear divide between the deployment of business skills within surveyors' general practice work and management consultancy roles. The issues considered in chapter 4 as well as this chapter, can still therefore be drawn on by surveyors undertaking mainstream general practice work. Indeed, there are around 11,000 members of the RICS Management Consultancy faculty and, as with all faculties, these represent elements of surveyors' work, and not necessarily the fact that surveyors specialise in a particular discipline, such as management consultancy.

Property rationalisation: various surveying issues

As an initial illustration of the need to understand the nature of a client's business, a general practice surveyor may be asked to advise a tenant on its property options following a reported decline in business by the tenant client. The tenant is occupying well-located, but not prime, 10,000 sq ft city centre offices. Two separate floors of a multi-let building are occupied, although staff numbers have fallen over the past six months. The rent passing is £120,000 (£12 per sq ft), with a rent review due in one year's time. The surveyor is to work closely with the business to determine the most favourable property solution.

Initial property investigations

The first property inspection may find staff spread thinly over two floors, with scope to consolidate onto one floor at minimal cost — thus improving operational efficiency for the business and securing cost savings through empty rates relief (and possibly other costs, such as business insurances and certain utility costs).

Leases are examined to establish the ease of relinquishing space if necessary (ie by assignment or subletting). Ideally, the agent acting for the tenant at lease commencement had the foresight to secure separate leases for each floor — as this allows one to be assigned and the other retained, whereas a single lease would usually prevent the assignment of part. Similarly, subletting may not be permitted for part of the demise. The surveyor will examine the time to lease expiry, any break provisions and any other important lease terms. Market research and comparable evidence shows market rents to be around £18 per sq ft, with the tenant facing a rent increase of £60,000 or more in one year's time (a 50% increase on the commencing rent from four years ago).

Business issues at lease commencement

The deal done at lease commencement is another illustration of the surveyor's need to consider a tenant client's longer-term issues and secure the necessary flexibility in leasing arrangements. The commercial imperatives of the tenant and/or a strong and competitive occupational market conditions may, however, have prevented the ideal lease terms being secured, with the landlord exploiting a favourable negotiating position.

Another business skills illustration regarding the initial letting is to prevent a client from entering a property and location that is likely to experience strong rental growth (ie localised rental growth ahead of broad market trends) unless such trends bring related benefits to the tenant. Future localised rental growth is likely, for example, in areas seeing a large number of new developments (or planning permissions being secured for future developments) but which have traditionally been regarded as secondary and low-quality areas, and command low property values as a result. A client's acquisition of freehold premises, rather than leasehold premises, in such circumstances, may be particularly astute (subject to the various other pros and cons of freehold against leasehold tenure).

Sensitivity of property costs to company profits

The surveyor would establish the need for the client to occupy offices in the subject location, and whether more secondary or even out-of-city space would be suitable. As rates, service charges and other costs are payable on top of the rent of centrally located offices, there may be scope to halve the rent and other costs and halve the space occupied — bringing a saving of around £125,000 per annum after reflecting the proposed rent increase.

The company accounts may, for example, show £3m turnover, £2.75m costs and £250,000 profit for the last financial year, and the savings in property costs could increase profitability by 50% from £250,000 to £375,000. However, an examination of the accounts may find that the two previous years' results were £4.5m turnover/£3.3m costs/£1.2m profit and £4m turnover/£3m costs/£1m profit. The business may be just struggling temporarily, and the £125,000 per annum potential cost savings, while appearing welcome, represents only around 10% of previous years' profits.

Businesses' reaction to downturn

Moreover, a more secondary or out-of-city location might impact on turnover, non-property costs, various operational aspects and, therefore, on profitability. The surveyor needs to work closely with the client business to judge the effect of a less attractive location on the ability to maintain existing business and win new business. There is a danger that the business reacts to declining profitability with a knee-jerk reaction of cutting costs in order to maintain profitability. However, this could simply serve to contract the business, making it harder to win new business and turn round what may have been only temporary difficulties. The relationship between fixed costs and variable costs, together with the size of profit margins, will also be relevant — such as a business with high fixed costs and slim profit margins suffering a more rapid downward spiral following losses in income/turnover than a business with low fixed costs and high profit margins (see p74 and p130 regarding costs, profit margins and such like).

Business location and impact on company turnover

In respect of location, a surveying consultancy, for example, typically needs to be centrally located in order to be accessible to clients and to create the right profile in terms of location and office specification. An insurance company, whose turnover relates mainly to telephone business, is unlikely to need well-located and relatively costly offices, and even if a ground retail/office outlet was needed centrally, the operations could be split. As indicated above though, if the difference in property costs between locations/property options is an insignificant proportion of a company's profit, the company might as well work from a good location (if simply only on the basis of everything to gain but nothing to lose). Location decisions and property costs can also be influenced by the preferences of the senior personnel who make property decisions, and which may be influenced by non-commercial factors. This is particularly the case when they themselves will be working from the property concerned and are conscious of the elevated personal standing considered to be bestowed by top-quality accommodation and their own grand office. Also, some buildings, especially headquarters, contribute to a company's brand — as demonstrated by the use of company's premises in websites, annual reports, promotional brochures etc (even when the building is multi-tenanted and the company occupies only part of it).

Business location and operational issues

With regard to the impact of a different location on non-property costs and the operation of elements of the business, a less attractive location may make it harder for the client business to retain and recruit staff. Many employees, for example, may now have to make two journeys to work — by public transport plus an undue walking distance — instead of the previously convenient single journey straight into the city. Amenities in the area, including retailing and leisure facilities, as well as free car parking on the premises, can attract staff. Also, a central location is conducive to team building and teamworking in view of its proximity of pubs in particular, whereas the relocation to more secondary or out-of-town offices may mean that evening activities are lost due to the lack of suitable outlets. A less together workforce can be more dispirited and less motivated. Location also impacts on logistics, such as the possible increased requirement, and therefore cost, of office cars.

Other accounting issues, costs and cash flow

As an example of a more detailed consideration, the surveyor needs to establish the ownership and structure of the business. For example, salaries and bonuses for its owners/principals included in the accounts as costs could result in a stated low profit when in fact, to its owners, the business is particularly lucrative. Further details on such accounting issues are given in chapter 7.

Relocation costs are also relevant (and the surveyor has to establish whether the business has any other premises). Consideration of the business' cash position is important, with property providing both scope to raise funds (such as sale and leaseback, or sale of a leasehold interest/assignment) and also drain funds (such as relocation costs and double overheads as part of relocation if the old lease cannot be easily assigned/sublet). As a further example of the surveyor's alertness to issues facing the business, if a lease assignment were contemplated, the surveyor would have to explain the possible ongoing liability for rent and other items beyond lease assignment if a new tenant, or successor tenants, fail to meet their obligations — ie privity of contract and the ongoing liability of the original tenant with pre-1 January 1996 leases, and liability under a possible authorised guarantee agreement with post-1 January 1996 leases. Existing leases can also take longer to relinquish than anticipated, especially in weak market conditions, and therefore adding to costs if the tenant has already vacated.

Surveyors as business advisers

There are, of course, a multitude of other issues facing businesses, but the above points illustrate some of the issues to be considered between the surveyor and the client. As indicated in the preface, surveyors can add to their mainstream surveying work by also providing business-related advice to clients — although the extent to which surveyors may take on such a role does, of course, depend on their areas of surveying practice and the nature of their clients.

The surveyor and the business may together conclude that the outlook for the business is reasonable, and they will remain in the current premises, seeking a subtenant for one floor, albeit on the basis that possession could be secured if need be, if and when the space is again needed for the business. The surveyor may also prompt the business' thoughts about home working, improved efficiencies through IT and more efficient space planning throughout the offices.

Agency

Surveyors' work in respect of agency, known also as property marketing, mainly covers lettings on behalf of landlords, sales on behalf of owners, leasehold acquisitions on behalf of tenants and freehold acquisitions on behalf of purchasers.

Market understanding

In the case of a letting or sale, the surveyor needs to evaluate the typical market for the property, and the type of occupiers that will be attracted. Sometimes this is relatively easy to narrow down, such as a prime high street retail outlet, which will inevitably be undertaken by a national retailer. Marketing can then be targeted relatively precisely, including through mailshots to retailers and retail agents (as well as site boards, marketing particulars etc).

Occupiers' operational needs

The agent still needs to understand the property needs of retailers, including individual retailers. There are certain requirements as to size, location and footfall, and if these are not met for an individual retailer, there is little point in distributing details to them (other than for courtesy notification, which serves wider marketing advantages).

The need to accurately understand clients' requirement is particularly pertinent when seeking to acquire property, such as an investment agent searching for suitable investment properties. Acquisitions will need to meet the investor's investment/acquisition criteria (such as lot size, location, tenant/covenant type) and be complementary to their existing portfolio. As clients require properties that meet their needs, agents who pass on details of all available investment properties do not come across well — wasting the client's time, and questioning whether the agent has understood the client's requirements. If there is considered to be benefit in keeping clients appraised of market conditions, other available properties, transactions etc, separate summary-style updates could be provided (or a suitable cover letter clarifies the "for information only" nature of the material).

Clients and their marketing budgets

When providing marketing advice, surveyors need to be conscious of their client's attitude to the marketing budget. Although certain interests can be let or sold with relatively minimal marketing cost, such as an off-market institutional investment sale simply introduced to a number of potential buyers, other interests will not have a defined and easily targeted market, and will incur greater cost.

Some clients will not appreciate the efforts and expense that is necessary to attract interest in certain properties, and to secure the best deal. Private clients may simply be reluctant to part with their money, and the public and corporate sectors can be constrained by marketing budgets. There are situations, for example, where an owner is anxious to secure a tenant of a £100,000pa property, but is troubled by marketing expense of £4000, which equates to only two weeks' lost rent — with considerably more weeks' lost rent being suffered because of the property's lack of exposure to the market. In contrast, successful investors and developers are well aware of the value that can be added by effective promotion, and can appear relatively lavish in the marketing initiatives deployed. The surveyor's knowledge of the client and the attitude to expenditure can therefore be key in ensuring that recommendations and courses of action are palatable.

Client management

It is important to prime a client against the possibility that a tenant or purchaser might not be found within the first round of marketing, and that further marketing expense may therefore be needed in the future. This is an example of how business skills extend to "client management". Other examples include the preference of presenting invoices to clients when successes have been achieved on their behalf, as opposed to including invoices alongside bad news, which can provoke a more adverse reaction as to whether advisers have provided good value. In addition, surveyors will not be seen in the best light by clients when advertisements placed in property journals and newspapers give undue prominence to the agent, and not the property and its attractions, by displaying large company logos and other information relating to the agent.

In-house issues and constraints

As an example of how in-house surveyors can deploy business skills in respect of their organisation's property marketing, consideration could be given to whether rental income and capital receipts are affected by budgets and any other internal accounting/administrative constraints. Here, year-on-year variations in property revenue can be seen as a natural occurrence in the property market, while relative obsession pervades with expenditure because of its greater scope to be measured and budgeted. Similarly, there can be no scope to incur expenditure on repairs or other improvements, which can enhance lettability or add value prior to sale, with properties being relatively free to remain on the market for some time, or simply sold in their current condition to the highest bidder.

Other illustrations of business and market awareness include the extent to which a small amount of expenditure on a property can have a disproportionately greater impact on its lettability/saleability and value. This includes basic cosmetic improvements to both residential and commercial property, which can maximise the feel-good factor on viewing. This is particularly important when typical applicants struggle to see possibilities beyond the property's current condition — in contrast with developers, for instance, whose focus will be on ultimate potential. The vision, confidence and ability to "spend a penny to make a pound" underlies much of business success.

Surveyors' procurement of suppliers/property advisers

Another business-related role of the in-house surveyor in respect of marketing, as with other fields of work, is to procure the services of external agents. In order to get the best from a property portfolio, agents' active involvement in a particular market, knowledge of likely tenants/purchasers, availability of databases etc, can be key. Again though, internal systems can sometimes restrict the scope to appoint the most preferable agent in terms of the potential for high performance, with minimum price being a prime consideration. When appointing advisers, public sector surveyors and other personnel can become anxious when putting other factors ahead of minimum price, reflecting perceptions of accountability and transparency. This also illustrates issues that private sector surveyors need to embrace when seeking to win businesses — including business models that facilitate low-cost services and price competitiveness through lower cost operations ahead of better-quality but higher-cost/uncompetitive options.

Business systems

Agency work is also a good example of where the internal running of a surveying business impacts on the success of work undertaken for clients — as opposed to simply influencing the efficiency and profitability of the surveying business. As the case surveyor will not always be directly contactable, systems need to be in place for others to send out details (such as particulars by post, or similar details immediately by e-mail) and answer basic queries. Consideration also needs to be given to factors such as the answerphone message provided and the availability of mobile numbers or colleagues who can assist. A poor impression is given, for example, when prospective tenants call on a Tuesday to hear the surveyor's recorded message regarding being out of the office or on the on phone on the previous Friday. Deals can be won and lost by such factors. Existing clients and prospective new clients may also be trying to contact the surveyor, and similarly be left with a poor impression. Surveyors need to ensure that reception and support staff systems operate smoothly, which includes maintaining a polite and professional telephone manner.

Presentational benefits and other attractions

A good agent not only understands the client's business issues well, but also can detect the principal needs of tenants or purchasers, and make the terms of letting or sale appear as attractive as possible. This extends to factors that can be drawn on by the tenant's/purchaser's agent or in-house representative in presenting a successful deal to their own clients or superiors.

The use of a rent-free period is a good example of how an investor/landlord can gain advantages in terms of rental income and capital value, if not immediate cash flow, while delighting a tenant with the terms on offer. A rent-free period helps a tenant's cash flow and can help flatter short-term profitability, but tenants are not always alert to the potential drawbacks. If, for example, a rent-free period of one year is granted in a 10-year lease with an upward-only rent review at year five, the rent may be £100,000 per annum (known as a "headline rent") — totalling £400,000 over five years. An alternative deal could be a "market rent" of £80,000 from day one, which is again £400,000 over five years. Some tenants think that a rent-free period is simply a concession, and are unaware that £100,000 is artificially high and not a market rent. Others tenants are aware of this, but still do not consider, for example, that if market rents are no higher than their current level in five years' time, £100,000, and not the market rent of £80,000, will be paid for in the next five years — ie a further £100,000 in rent is paid over five years. From a presentational perspective, a rent-free period simply "sounds good". Where surveyors act for tenants though, it will important to highlight wider business and budgeting issues in respect of rent-free periods and any other potentially adverse elements of a deal.

Property management

Terms such as property management, estate management and landlord and tenant are used in different ways in practice, but broadly include rent reviews, lease renewal, assignments, sublettings, dilapidations, service charge management, insurance, buildings management and such like.

As with agency work, it will be important to embrace clients' issues and objectives in the advice provided, including any strategies recommended.

Assessing clients' objectives: investment illustration

Property investors generally seek income/rent, income/rent security, income/rental growth and capital growth — although, as indicated later in the chapter, different investors have different lead objectives and business plans. In terms of the basics of agency and property management work, this means that surveyors need to consider the impact of lettings, rent reviews etc, on a client investor's wider objectives.

The level of rent achieved at review, for example, will influence the capital value of the property — which will be important to the client if wishing to sell or raise finance. Other factors include the benefit of higher property values in helping prop up a company's balance sheet and general borrowing capacity, or the benefit to a fund manager of strong performance figures.

In some cases, the rent review may have to be accelerated in order to meet such objectives in a certain timeframe. On other occasions, tactics may be influenced more by property factors than business factors, such as to delay the rent review pending better comparable rental evidence. Even then, another business dimension will be the scope to achieve favourable settlements on some units, and for this to be used as beneficial evidence on other units — again helping to influence capital value.

In contrast, any delay in concluding a lease renewal could be damaging to a client's investment interests, albeit temporarily, owing to the uncertainty as to whether the tenant will remain, and the impact this could have on both capital value and cash flow. A client investor's cash flow position could be another key influence on the course of action recommended with any property dealings.

Pro-actively adding value

One option open to the surveyor in avoiding the above investor's business disadvantages around lease renewal is to seek a surrender and renewal of the existing lease some time before lease expiry. This is an example of "asset management", which generally means pro-actively looking for opportunities to enhance value (whereas estate management concerns the general addressing of ongoing issues).

A refurbishment could enhance both rental and capital value considerably, but the investor will have to carry a negative cash flow for such a project — with refurbishment, possibly finance/interest repayments, and other costs being incurred while no rent is being received. Here, the surveyor's understanding of the rest of a portfolio will be important, including rental income, all outgoings, the scope to refinance properties and the scope to sell properties and realise funds if need be. Economic and property market conditions will need to be evaluated and, as a general business point, "worst-case scenario", as part of a range of risks, will have to be contemplated.

Other examples of asset management include changes of use — not simply alternative/development uses for generating favourable sales receipts, but also changes of use such as ground floor offices to retail use, or changes in certain types of retail uses to other uses, both of which could enhance rental and capital value.

Business-property illustration: shopping centre management

The management of a shopping centre is a good example of where the surveyor needs to understand retailing fundamentals, and how different types of use and different occupiers contribute to the footfall and success of the centre.

The total footfall and the type of footfall influence the money spent by customers throughout the centre, the profitability of retailers, the demand for space, the rental levels achievable, the capital value of the property and the investment returns realised. The term "loss leader" is not common in property, unlike certain other areas of business, but an example would be to include a foodcourt in a shopping centre, even though it generates considerably less rent than a retailer would pay. Such a facility ensures that shoppers can obtain food and refreshments, and stay in the shopping centre (and spend more money) instead of having to seek facilities elsewhere. This also adds to customers' shopping experience, and helps ensure loyalty to the centre — rather than shopping elsewhere. The centre's ability to provide all typical shopping requirements (and "tenant mix") also helps to concentrate spending in the centre. Proximity to public transport, car parks, other attractions and the development of new retail space in the area are further examples of property considerations.

The essential interrelationship between the surveyor's business and property knowledge is that to be able to remove and replace tenants, leases need to be sufficiently short, or break clauses included, with "contracting out" of the security of tenure provisions of the Landlord and Tenant Act 1954 also being important.

In contrast to shopping centre management, the actual trading success and profitability of tenants in offices and industrial properties is of relatively little concern (other than their financial standing/covenant strength needing to be acceptable).

Valuation

Business issues are not reflected in valuation work as significantly as the aforementioned property dealings, but the valuation of specialised property interests warrants a detailed understanding of particular markets and industries.

Understanding specialist trades

This is primarily in respect of the licensed and leisure, or "trading", properties whose value is dependent on the profitability of the venture that can be run from the premises. Examples include pubs, hotels, nightclubs, casinos, petrol stations and care homes.

Valuation methodology will often involve capitalising a profit figure by a multiplier in a similar way to rents being capitalised in the case of investment valuation. As well as understanding typical industry issues and their effect on value (such as licensing laws, health and safety legislation), surveyors need to understand the components of business profitability. It is also important to distinguish between profitability that could be achieved by the market at large (and would be reflected in a sale price/valuation achievable) and profitability that is dependent on the current operator (and would not relate to an open market scenario). A national hotelier or pub chain, for example, may benefit from special branding, key marketing/business concepts and considerable goodwill, securing profits that are highly unlikely to be achieved by others. In contrast, a small pub owner may not have the ability, personal interest or finances to exploit the property's true potential. It will be particularly important for surveyors to be familiar with industry norms, such as occupancy rates in hotels, profit margins from pubs and restaurants and staff costs in the healthcare sector. Such factors will also be crucial in other property dealings in specialist property types.

Valuation and the corporate sector

There is a particular link between property valuation and major events in the corporate sector — such as high-profile takeover bids and the collapse of well-known companies. In the supermarket sector, a 2003 takeover bid by Morrisons of Safeway soon saw attention turn to the valuation of property assets — in particular, the true value of Safeway's property holdings and whether the bid fairly reflected this. Similarly, part of Marks & Spencer's defence to Philip Green's takeover bid in 2004 was a revaluation of property assets. Indeed, unduly low asset values, which help a share price to remain relatively low, can make a company a more attractive takeover target — even to the extent that companies will be asset stripped in whole or part in order to realise maximum value.

In contrast, some companies will be content with lower asset values as it helps flatter their return on capital employed. Other situations see investors, developers and also the corporate sector looking for the highest possible capital values in order to prop up balance sheets, improve borrowing capacity through greater security, enhance share prices through higher net asset value and/or also contribute to performance analysis. Chapter 7 provides more detail on this.

Business issues and professional ethics

Such accounting and financial factors illustrate the importance of property to businesses, and the issues in which surveyors may be involved — sometimes for the benefit of the businesses concerned, and on other occasions, to provide valuations that suit certain purposes. The problems of the Queens Moat Houses hotel chain in the early 1990s was due largely to the overvaluation of its property assets and consequent extensive writedown, with the Department of Trade and Industry, after many years of investigations, criticising the accountancy and valuation practices adopted and questioning the valuation methodology of the surveying profession. Changes to the RICS Appraisal and Valuation Standards (the Red Book) in 2003 partly reflected the need for the profession to withstand the impact of scandals, such as Enron, through the activities of valuers. A wider business issue for property professionals is, of course, the integrity demonstrated by the profession to its markets.

Local authority property management

By considering more closely the issues in particular sectors, an examination of local authority property management also illustrates the difference between client dealings in the public and private sector. The key to delivering appropriate property solutions, as shown below, can be found by carefully considering clients' business objectives and other operational issues.

Local authority property portfolios

Local authorities' property assets range from major office and retail developments in prime city centre locations to strips of land left over from road-widening schemes. This warrants a varied management approach, maximising rental income and sales receipts from some properties and minimising liabilities and maintenance costs on others.

Local authority property portfolios vary throughout the country; while some comprise mainly properties occupied by council services, others include extensive investment interests, major development sites and council housing stocks.

The majority of local authorities' property interests are held in freehold ownership as owner-occupier or landlord, but some operational properties will be held on a leasehold basis. This involves the usual estate management functions of lettings, rent reviews, lease renewals, service charge management and rating.

Many local authorities, as with other public sector property teams, handle the majority of case work themselves and appoint external consultants where particular expertise is required (or on special projects for resource reasons). However, some local authorities have a predominantly procurement role, outsourcing the majority of property work to external consultants.

Performance monitoring

Local authority property teams, as well as others in the public sector, are under ever-increasing pressure to run their property interests effectively. This warrants the implementation of suitable property management strategies and ongoing monitoring against performance criteria.

Local authorities' property work is governed by the Office of the Deputy Prime Minister (ODPM). In addition to the general local authority reporting procedures in place, a corporate asset management plan is prepared, with various statistics being required by government on an ongoing basis.

Local authorities also maintain statistics to support their overall role as property managers and to help specific functions. Examples include total book value of non-operational assets (split between development/sale interests and investment/income producing interests) and rental income and rental voids as a percentage of investment portfolio value. Internal rates of return for the portfolio can be compared with other councils, and private market indices, although it is important to consider the make up of a council's non-operational portfolio when making comparisons.

Traditionally, local authorities have not adopted asset management plans involving investment expenditure, the acquisition of adjoining sites and surrender and renewal transactions, but the government is encouraging a more pro-active approach. This could include buying properties for investment, particularly those also contributing to the council's overall economic development objectives.

Best value

"Best value" involves the requirement of local authorities and other public bodies to competitively tender work to contractors and consultants, for example, with larger value work warranting particularly stringent processes. External surveying practices seeking council work usually bid for work alongside two or more other practices although, depending on the nature and value of the work, some appointments come direct from an assembled panel of practices, working to pre-agreed rates. In contrast, the private sector is often at liberty to appoint an adviser of choice, at whatever remuneration, without an obligation to tender (although rules may become more stringent for larger concerns).

To win work from the public sector, it is important to demonstrate an expertise and a strong track record in specific areas, have suitable policies in place that share the public sector's approach to equal opportunities/anti-discrimination and be cost-competitive. Notwithstanding the orientation to the lowest price, reflecting accountability and transparency, in suitable circumstances, wider criteria are still adopted in the appointment of advisers.

Best consideration — property transactions

"Best consideration" refers to the need for property transactions to be the best commercial deal. Properties for letting and sale have to be marketed with a realistic budget, also adopting the correct advertising outlets and, in the case of sale, the appropriate method of disposal. Local authority properties are often sold by tender, whereas the private sector may often adopt private treaty. Private property owners have greater flexibility to accept quick and/or cash deals for their commercial advantages and are unconstrained by issues of accountability and transparency, which are associated with public assets and taxpayers' money. Auction disposal is transparent, but it is important that only the right properties are included, and best consideration will be achieved for each of them.

Best consideration does not necessarily mean best price, although there need to be strong reasons for accepting a tender that was not the highest price. A council's need for immediate receipts would be unacceptable but if, for example, a bidder had a track record of withdrawing from transactions and/or he had to raise finance on the property to be purchased, which was not in fact mortgageable, a deal could be negotiated with a second bidder who is better able to complete within a reasonable timescale.

Occasionally, sales will take place below market value — for example, in the cases of social provision, a council's work with charities or in connection with regeneration areas, where individual sites contribute to wider objectives and revenue opportunities. Special approvals will be required, usually through internal council processes, but sometimes from the ODPM.

For some public sector interests, independent valuations are sought from external consultants in order to support decisions to sell property (and similarly with any purchases).

Internal client relationships

Local authority property teams are really agents to their client council (such as in respect of non-operational/surplus/investment property) and to individual client departments (such as in respect of an education department's operational property interests). This involves similar reporting requirements as in private practice, albeit to a departmental client. Advice may be required by a council's finance team in respect of capital and rental revenue projections. Reports may also be prepared for council committees. Although the majority of property transactions will be conducted on the usual commercial basis, some property issues have a political dimension.

The public sector as clients

In the same way that in-house public sector representatives embrace the concept of best consideration, political issues etc in property dealings, the advice of surveyors similarly needs to do so. Property advisers to the public sector should not rely on the client's superior understanding and expertise of such issues, but should understand the detailed implications themselves. As with all advisory roles, clients are looking for options, advice, decisions and recommendations. They are ideally placed in a position to agree a recommendation, rather than make decisions themselves, or have to respond with further information because of a property adviser's indeterminate stance. Clients also wish to avoid undue time in serving an external instruction through time incurred by their own staff on matters that the property adviser should be able to progress without undue liaison. For example, the outsourcing of instructions to property advisers may be administered on a day-to-day level by junior and/or non-property employees, who need to take property advisers' queries on matters such as best consideration and other public sector issues to a higher level, thus causing client frustration and delays in progressing the case work.

RICS Policy Unit research

In 2002, reflecting the increased use of local authority property by third parties, and government considering the extent of property being used as a form of subsidy, RICS, together with the University of Reading, published a research report entitled *Whose property is it anyway?*

A key point highlighted was the need for robust arrangements for dealing with third parties, including formal lease/licence agreements, charging of appropriate rental levels and accounting appropriately for any cross-subsidies. The report also highlighted the scope for efficiencies to be gained through the inter-departmental co-ordination of accommodation requirements and the commercial use of appropriate surplus space by third parties. It is not uncommon for local authority property teams to be unaware of certain property dealings taking place by council departments, and the research highlights the importance of local authority property teams ensuring that managers across the authority are aware of the property issues.

RICS has recently produced *Asset Management in Local Authorities*, which provides guidance on how local authorities can produce asset management plans for their property assets. This includes reference to the various other public sector reviews and initiatives (such as the *Gershon Review* and the *Prudential Code* — the details of which are beyond the scope of *Business Skills for General Practice Surveyors*). Details of the guidance are available from RICS at *www.rics.org/Property /PropertyManagement/AssetManagementinLocalAuthorities230305*. In chapter 4, the section on the Lyon's Review (p63) further illustrates of the combination of property, economic, political and demographic issues in the public sector.

Regeneration

The regeneration sector is another good example of the importance of surveyors understanding clients' strategic objectives, and how these relate to individual property dealings.

English Partnerships is the national regeneration agency, and there are also regional development agencies (RDAs) throughout the country, such as Advantage West Midlands, East Midlands Development Agency and Yorkshire Forward. Also, various smaller, sometimes project-specific regeneration companies, as well as many local authorities, are involved in particular regeneration initiatives.

Understanding an agency's business and regeneration objectives

The purpose of regeneration is to support economically disadvantaged areas and generally help contribute to economic growth. This is achieved by many means, including business grants and training schemes, and also property dealings — with the acquisition and development of sites often being key to delivering regeneration objectives.

Whereas private sector development is usually orientated to the higher value development opportunities having higher development land values, regeneration activity, by definition, is orientated to the more disadvantaged areas that have low values. Development/alternative use values will often be negative, simply because the value of the completed development is insufficient to cover land acquisition costs (at existing use value), construction costs and the developer's required profit. Construction cost often includes the remediation of contamination and provision of infrastructure, both of which can often be extensive. Because of such factors, a regeneration agency will need to provide grant funding in order to ensure that physical development actually takes place (at least initially — see "pump-priming" below).

Surveyors working in regeneration need to reflect their agency's economic objectives within property dealings. Property acquisitions and grant funding to developers, for example, need to reflect key criteria, such as the re-use of brownfield land and the number and type of jobs created. Acquisitions and expenditure to create new uses will also need to be a "strategic fit" with the agency's corporate plan/economic strategy for the region (or nationally in the case of English Partnerships). This is also in line with government policy directives and regional and local planning issues.

Understanding sector characteristics and market trends

Surveyors also need to understand how early property development in a regeneration scheme contributes to the uplift in values of surrounding land. Key to the ultimate success of a scheme will be the attraction of private developers and investors, and the speed at which regeneration takes place within budget constraints. As development becomes commercially viable because of increasing values, and private developers can profit similarly to their ventures in non-regeneration areas, more development becomes led by the private sector. This process is known as "pump-priming" and is why regeneration agencies are sometimes described as having a catalytic effect on an area. However, notwithstanding the willingness of the private sector to become involved without requiring grant funding, regeneration agencies still take an active role in securing the right uses for an area. It is also necessary to ensure that development takes place and developers do not, for example, landbank sites pending further increases in value, thus leaving undeveloped pockets of land that detract from the overall scheme being pursued.

Understanding business issues: inward investment

Another key function in which surveyors may be involved, and which incorporates mainstream business skills alongside the understanding of clients' objectives, is inward investment. This involves the attraction of overseas companies to the UK, although the term is sometimes also used, for example, when companies are attracted from outside a regional development agency's own region. In a regeneration scenario, the investment courted will often be in a particular sector, reflecting the uses being promoted in an area, and which are compatible to the agency's economic objectives — such as a particular type of employment that can easily displace lost jobs in a declining industry, or which can spurn the growth of surrounding supplier industries. In contrast, the private developer or investor typically seeks the best tenants, best rents and highest end/sale value — albeit conscious of the effect that larger occupiers (including "anchor tenants") can have on the ability to attract other occupiers, and their consequent ability to secure rent and/or other concessions.

The surveyor needs to understand the requirements of prospective inward investors and be able to demonstrate how these can be met with the available property and within the local economy. As with most relocation and business investment decisions, examples include the availability and adequate supply of suitably skilled labour, proximity to the business' markets/customer base and transport/accessibility factors.

Anticipating wider effects, reading markets

As an illustration of the wider benefits of regeneration, the funding of construction, as with the majority of government/public spending, supports employment in the construction sector and the supply sectors. Another example of such an effect is when a car plant is mooted for closure, and not only its own employee's jobs are in jeopardy, but also those of local components suppliers. This reduces consumer spending power within local economies, and also has property market effects of lower house prices and falling industrial rents. Business skills include the observation of such chain reactions, in the same way that in chapter 2 examples were provided of the importance of surveyors' ability to read markets.

Public relations and marketing

The regeneration sector also provides a good initial example of the importance of surveyors' alertness to public relations. The work of regeneration agencies is occasionally criticised by communities, developers, unsuccessful applicants for grant monies and even government and other bodies. Because planning, compulsory purchase, land assembly, construction etc takes time, all within budget constraints, it can appear to non-property people that little is being achieved. European state aid funding rules, landlord and tenant rights, and the need to relocate business also hinder the pace of progress. However, once developments come on stream and the more visible signs of success are apparent, the agency's image is presented more favourably. The press can perceive that an agency's performance has improved but, in reality, people, processes and performance have remained unchanged.

Surveyors working for organisations that have an essential public profile to preserve, need to work closely with the public relations department. The image of the organisation can be important in attracting developers and investors to regeneration areas, and any reported problems in the press need to be balanced with success stories. Press releases minimise the chance of inaccurate reporting, as well as making journalists' jobs easier than having to search for a story. In addition, an arrangement to feed exclusives to particular press outlets and journalists can lessen the chance of adverse coverage.

Property developers need a good reputation in order to increase the chance of winning planning permissions, increase agents' willingness to put properties to them and increase property owners' willingness to pursue transactions (noting that some developers and investors create a poor reputation in individual dealings, which can have detrimental wider effects). In the case of the corporate concern with large property holdings (such as a utility), the marketing of property and the brand of business are interrelated. A corporate retail occupier looking to expand its outlets may offer imaginative incentives to surveyors introducing new properties. Property teams within local authorities may need to create the right presence among community groups. For all businesses, the handling of telephone calls, the speed of e-mail response, the quality of the office environment and many other factors effects the image portrayed to others.

For property consultancies, such factors are pursuant to business development — maintaining existing business and winning new clients and new instructions. Marketing also means developing the right brand and being alert to the image of the consultancy within the market and in the eyes of its clients.

Property development and property investment

Property developers and property investors generally aim to make as much money as possible, within acceptable risk tolerances.

Surveyors tend to be familiar with property valuation methods, such as the residual method for development interests and the investment method for investment properties, but do not readily see how developers and investors look to make their money, unless working for such firms on an in-house basis. Even then, it can sometimes only be at senior level, where knowledge of the detailed economics of the business is accessible.

To get the best out of a property development or investment business, an understanding is needed not only of the inputs into valuations and appraisals, but also how market conditions and many other factors might change — and how this can create opportunities, but also present risk. Surveyors who advise developers and investors are highly valued when they can similarly reflect such factors in advisory and transactional work — and as with any client, embrace their business objectives, cultures and working methods.

This section provides basic illustrations of how development and investment returns can be influenced by changes in economic, occupational/rental market and investment market conditions, and how finance/gearing presents advantages but also risks. This is part of a number of illustrations within *Business Skills for General Practice Surveyors* that shows the sensitivities relating to the costs of a business (including fixed costs and variable costs), profit margins and other aspects — all of which again influence opportunity and risk. Good business people instinctively picture such factors as part of their evaluation of opportunities and risks, and as part of their decision-making.

Property development

Set out below is a simplified illustration of a development appraisal of a 50,000 sq ft office development in a city centre. The developer is looking to acquire the site, commence construction, find tenants and sell the completed, well-let, investment to an investor.

Market rents are £20 per sq ft, and market yields are 7%, producing a rental of £1m and a gross development value of £14.3m (£1m, Years' Purchase (YP) in perpetuity at 7% – multiplier 14.3).

Gross Development Value (GDV)		£14.3m
Build costs	£9m	
Developer's profit	£1.5m	£10.5m
Land value/purchase price		£3.8m

From a valuation perspective, the surveyor has to be satisfied that the inputs are accurate and the approach mirrors that of developers in the market at large. A suitably cautious approach has to be taken and regard given to the precise definition/assumptions relating to Market Value (MV) in the Red Book. With an appraisal on behalf of a developer, inputs similarly need to be accurate, but having regard to issues specific to the individual developer. Examples include the ability to build particularly cheaply, the lack of expertise regarding contamination remediation and therefore greater caution through an increased build cost contingency, the requirement for a relatively high developer's profit because of the number of possible schemes that could be pursued, or the acceptance of a relatively low developer's profit because there are no other opportunities available (and land acquisition is essential in order to maintain development activity, and retain an in-house workforce/contractors).

Moreover, the developer is considering risks and opportunities. With a speculative development, if the developer was to acquire the site at £3.8m, and there was a construction period of 18 months, the rent and the yield could change prior to tenants being found and the property being sold. A 20% fall in rents to £16 per sq ft would lead to a loss of £1.4m (site £3.8m, build costs constant at £9m, GDV £11.4m). If, at the same time, yields increased to 8.5%, losses would be £3.4m (site £3.8m, build costs £9m, GDV £9.4m). Similarly, a 15% rise in rents to £23 per sq ft would result in an increase in developer's profit from £1.5m to £3.6m (site £3.8m, build costs £9m, GDV £16.4) — in other words, profit more than doubles. If at the same time yields fall to 6%, the developer makes a profit of £6.4m (site £3.8m, build costs £9m, GDV £19.2m), which is more than four times the profit required at the outset.

The smaller the developer's profit requirement (ie the percentage required of build costs, or of GDV), the more sensitive the effect of changes in rents, yields and also build costs will be on the total profit earned. This is the same principle as "profit margins" for non-property companies.

These illustrations show the success of property development is largely contributed by the performance of the underlying markets — such as offices, retail, industrial, residential and investment. This highlights the importance of surveyors being able to understand property market trends and judge, as best as is possible, the outlook for relevant markets. As is outlined in chapter 5, economic factors play an important part.

The illustrations perhaps explain why certain developers consider a "fag packet" valuation/appraisal adequate (and why little regard may be given to some of the finer academic principles behind the process). Importantly, developers are looking for the right markets and opportunities to exploit. Sometimes today's rents, yields, costs etc mean that development is not commercially viable, but it is nevertheless pursued because of the beneficial effect of anticipated market changes.

The example also shows how in a market with a strong outlook, surveyors can price properties based on current values, only to see developers bid considerably higher figures, as they strongly believe that certain markets will rise. Conversely, at the top of the market, land values can remain too high for too long, with transactions slowing because developers become concerned about the effect of interest rate rises, or the weakness in property markets and, therefore, are unable to acquire sites at the right price.

There are, of course, many more issues to consider as part of property development in practice, including the need to build to cost, build within timescale, the requirement to attract tenants and the pros and cons of pre-lets and financing arrangements. Success in dealing well with all such factors is underpinned by a range of business skills (see "Business skills summary", p42).

Property investment

To illustrate the opportunities and risks associated with property investment, the position of the investor reliant on finance is considered. As demonstrated, this creates sensitivities that, as with the development illustrations, mean more money is made when things go well, but can result in the collapse of ventures if things go badly.

When outlining the role of lenders on p14, an example was provided of a property investor making a £4m acquisition, with the investor borrowing £3m (debt) from a bank and committing their own funds of £1m (equity). If the property yields 8% rental income (£320,000pa) and benefits from an increase in capital value in the first year of 10% (£400,000), this represents a total return of 18% on £4m. However, as the investor has committed only £1m, at a loan to value ratio (LTV) of 75%, if finance is available at 6%, the investor makes a total return of 54% on their £1m outlay. This comprises £320,000 rent, less finance cost/interest of £180,000 (6% of £300,000) = £140,000, add £400,000 capital increase = £540,000, which divided by £1m = 54%. (Although the calculations on p14 were undertaken slightly differently, they produce the same figures.)

If following acquisition, rents fall to £250,000 as market conditions deteriorate and some tenants leave, and capital values fall by 20% for the same reasons, the net rental return after finance at the same 6% is £70,000 (£250,000 rent less £180,000 interest). The capital loss is £800,000 (20% of £4m) — ie leaving £200,000 equity of the initial equity investment of £1m. Cash flow is also becoming precarious, with loan repayments as well as interest payments having to be met (unless interest-only facilities are in place). Acquisition and management costs would also have to be considered in the above example, but there would be scope for increases in rental value over time.

The surveyor needs to remain alert to the many factors affecting such opportunities and risks. The case example of the family property business below provides more illustrations, and an example of an investment company's accounts in chapter 7 shows the importance of surveyors' understanding of accounting and financial matters in order to help clients/investors get the best out of their businesses.

Case example: family property investment business

The following case example concerns a family property investment business in a large UK city. Properties in its £8m portfolio range from £40,000 studio flats to £400,000 multi-let mixed residential-commercial interests, including student lettings. Gross rental income is £1.5m.

Family members are the only directors, and are also the majority shareholders of the company's £3m equity/shareholders' funds, with the remaining shareholdings also being family-based. Full-time staff are mainly family members and management advice is provided by a consultant chartered surveyor.

Contributors to rapid expansion over recent years include the injection of fresh equity/capital from shareholders, equity within property released on remortgaging following increases in property values, the provision of bridging finance and other temporary loans, the support of three lenders and the effective management of high gearing and a sometimes precarious cash position. Strong profitability reflects the high-yield nature of the portfolio/sector, the intensive management and the family's commitment.

The company needs to consider its options in respect of further expansion, bearing in mind that the current shareholders do not have any new equity to inject. The majority shareholders are apprehensive about a reduction in family control, but acknowledge that this is a prerequisite for the next stage of growth and for realising, at some stage, their original aspiration of stock market flotation. The principal two options available to the company are outlined by the consultant surveyor as follows:

- **Option 1: manage and maintain existing portfolio**
  This confines the foreseeable future to the management of the current portfolio without seeking equity funds from new shareholders/investors. Although acquisitions will be possible from profits, generally growth will be slow. Furthermore, as lenders are generally reluctant to support relatively small single acquisitions, it may be necessary to progressively acquire properties individually for cash in order, in due course, to mortgage a number of properties.

  Advantages of this option lie in the low risk profile and the reassurance to the family shareholders that they have full control. Disadvantages lie in the loss of momentum to growth, reduced motivation and time commitment of current personnel brought about by the absence of new challenges and daily pressures, the lack of regular work to retain a predominantly in-house workforce of both skilled contractors and casual labour, and the difficulty in attracting quality professional support (from a consultant surveyor or other professionals looking to gain from a share of the equity, and share options, for example, in addition to fees).

- **Option 2: relatively rapid expansion, loss of majority shareholder/family control**
  If further rapid expansion is to take place, new sources of equity will be needed, which, given the current shareholders' lack of new funds, would have to be sought from other shareholders/investors, possibly including additional directors and external consultants. New staff would have to be attracted and key personnel offered satisfactory remuneration.

  Advantages include the potential wealth that can be realised, extending beyond the retirement of the current shareholders/directors, who may otherwise be difficult to replace if the business took option 1 of mainly managing the current portfolio.

  The company is now well positioned to enjoy further strong expansion, subject to a short-term improvement in its cash flow and profits position and to minor refinements in its management systems. Such a move, from an essentially family enterprise to a company serving all shareholders equally, may, however, require too substantial a change in approach to be acceptable to the existing majority/family shareholders, especially as loss of majority family control would be likely.

  Consideration needs to be given to factors such as the extent to which majority shareholders are comfortable with the involvement and requirements of new shareholders/directors, the extent to which other investors actually wish to risk committing equity to an essentially family business and the overall aspirations of the company for growth. Future options include the sale of the company to a single investor, assembling smaller portfolios for sale, piecemeal sales of single properties or stock market flotation.

  Another consideration is the possible buyout by the main family shareholders/directors of the other family shareholders not involved in the business. This may be beneficial from a long-term management and expansion perspective, despite any available funds possibly being better used for acquisitions in the short term.

The way forward

After the company decided to remain as a family-based enterprise in the near future, the consultant surveyor outlined some of the opportunities and the risks it faced, together with guidance on how these could best be managed.

The advice would have reflected the change of the company's investment profile from high gearing and an often-restricted cash position, to a lower-risk profile providing a more cautious approach to growth, with less reliance on external professional support for day-to-day matters.

Key issues for the business

The key issues outlined by the surveyor in taking the business forward are as follows: combining factors of strategic relevance with the day-to-day issues that would be critical to the company's success.

- **Investment and growth strategy**

  As mentioned above, acquisitions, with limited new equity, will be gradual and growth generally slow. Nevertheless, there are some properties within the portfolio that could be remortgaged, thus releasing equity for new acquisitions. In carrying properties at less than full value, however, a further cushion is provided against risk and adversities.

  When acquiring new properties, consideration needs to be given to the usual factors, such as property type, percentage rental returns, management issues, extent of refurbishment expenditure, location, geographical diversity, scope for short-term capital gains (occasionally for sale but mainly for refinancing) and balance within the portfolio.

  Acquisitions, refurbishment and maintenance need to be timed in relation to the regular employment of any retained in-house workforce, and to the overall cash position. A vacant property requiring refurbishment is incurring interest repayments and works expenditure, while not receiving rent.

  A keen eye must be kept on economic and property market fortunes, internationally, nationally and locally — all of which can affect interest rates, rental demand, rental levels, capital values, profitability and so on. Changes in landlord and tenant law and planning legislation/guidance can similarly have an impact.

- **Cash position**

  The company must aim to be in a positive cash position at all times, avoiding reliance on an overdraft facility. This will provide a cushion against periodic cash flow difficulties, as overdraft facilities should be drawn upon only in essential circumstances.

The student residential sector accounts for around 30% of the company's income. As most rents are received termly in advance, cash flow planning is vital. It is especially important to account for voids and half-rents during the summer months, noting that once termly cheques are received in April, full rents are not received again until October.

The company must plan for interest and loan repayments due to lenders. Consideration must be given to the point at which loan repayments are made on loans that are currently interest-only, and to the point at which fixed-interest repayments for some properties revert to a possibly higher variable rate. The level of interest rates does, of course, need to be monitored, in terms of both the Bank of England base rate and the competitiveness of rates set by the company's lenders. It is essential to plan for other liabilities, such as taxation and major works.

The personal finances of the majority family shareholders actively involved in the business as directors are also ideally structured to provide further guard against temporary adversities. Profitable businesses can still fail because of poor cash flow. If the company fails, losses would be likely to extend beyond the company to directors individually, owing to the personal guarantees provided to lenders and others.

- **Lenders and investors**
  It is important to work with a number of lenders, and therefore ensure that none can exert undue pressure in respect of interest rate and other charges, or the company's overall operation. New lending arrangements need to be structured to fit with existing arrangements, portfolio issues and expansion plans. At all times, consideration must be given to the effect of certain courses of action on the company's quarterly management and full year-end accounts, which will be presented to lenders and any new investors in the future.

- **Personnel and professional**
  The company's in-house personnel, comprising family members, must have the expertise to deal with day-to-day matters, but also must recognise when external support is required.

  (Family enterprises are generally a specialist area of business consultancy owing, for example, to issues of maintaining family control, slowness to embrace change, problems with any poorly performing and/or dissenting family members, the attitude to outsiders and the need to plan for succession.)

Property-specific issues

The surveyor, as business consultant to a small property investment company, should be easily able to advise on property-specific issues, such as marketing methods and agency law; property management/IT systems; property purchases and sales; the terms of residential tenancies and commercial leases; rent reviews and lease renewals; minimising rates liabilities; administering service charges and rent collection; dealing with covenant breaches; complying with health and safety laws; ensuring planning consents are in place and arranging insurance.

The above example is based on work undertaken by a surveyor (and was also featured in the main *Estates Gazette* journal). The company accounts of one of the surveyor's other investment company clients are shown in chapter 7 together with further illustrations of how surveyors can combine property knowledge with economic, financial, accounting, managerial and general business acumen.

Telecoms

Telecoms is a good example of a specialist property type that needs specialist skills from surveyors who understand the operational nature of telecoms operators' businesses. The extent of telecoms operators' onsite telecommunications equipment, for example, influences the rental value of the site, and surveyors without the requisite technical understanding cannot act effectively for clients. It is also an example of how markets can experience rapid change, in terms of both the market for property interests and the market for professional/surveying consultancy services. This is illustrated in the following extract from *Starting and Developing a Surveying Business*.

> The telecoms operators' need in the early 2000s for property acquisitions to accommodate telecoms equipment and secure coverage for their networks helped provide strong earnings for those surveyors/practices having entered the market in anticipation (with a business plan recognising the likely short-term potential of the market). Operators and therefore surveyors' acquisition activity inevitably later subsided, and although increased telecoms property holdings created more property management work, more surveyors/practices had entered the sector. When the commercial imperatives of the operators were to roll out their networks, fees paid to property consultancies would be a relatively small consideration — especially as a consultant's expertise and profile could influence whether a planning consent was won or lost for a site. But once the operators began to manage an estate comprising numerous low rental interests, in a more competitive market for surveyors/property consultancies with expertise in telecoms, the profitability to the surveyor/property consultancy became more limited. Potential instructions also diminish if an operator, like any large client, decides to undertake work in-house through an expanded property team.

Starting and Developing a Surveying Business also comments that surveyors are unlikely to seek to get in and out of a particular market, and "make their money and run" in the same way that retail and leisure operators may exploit fashions and trends, but the astute surveyor still searches for such opportunities. Recently, access audits pursuant to Part III of the Disability Discrimination Act 1995, effective from October 2004, and asbestos regulations, effective from May 2004, have provided surges of profitable work to those surveyors/practices suitably positioning themselves in advance.

Business skills summary

Within the aforementioned examples of surveyors' general practice surveying work, a range of business skills has been highlighted.

Understanding clients' business issues

A common theme has been the need to accurately consider a particular client's operational business issues and objectives, and reflect these in the advice provided. An understanding of occupational markets (retail, office, industrial), investment and development markets, and the economy has been shown. As well as stock market performance being integral to economic issues, share price performance will be important for some clients, and property dealings will reflect this. Accounting issues have been highlighted within this chapter, and are illustrated in chapter 7 — ranging from an understanding of property costs in relation to a business' other costs and profit margins, to accounting and financial matters being a key driver in a business' profitability. Chapter 4, in respect of management consultancy work, also demonstrates the need for surveyors to understand accounting and related business issues.

Softer skills

There is a different emphasis among surveyors on the range of softer business skills needed, but negotiation techniques are essential in most areas of general practice surveying. Presentation techniques are needed in respect of businesses pitches by property consultancies and all surveyors will, at some time, have to give talks — ranging from team meetings to presentations to over 100 people. Interpersonal skills and the ability to communicate effectively are vital in many areas of business. Skills such as assertiveness and an ability to influence people are also needed. The quality of reports and other written communication needs to be high in order to convey credibility and a grasp of the issues, as well the more subtle side of being synonymous with high-quality service/technical knowledge. Many other softer business skills include listening (as opposed to interrupting and/or talking over others), chairing meetings (which retain focus and do not become talk-shops on side issues), social skills (helping build good working relationships) and team skills (as opposed to appearing too single-minded/self-centred).

Management skills

Younger surveyors will, in due course, hope to become managers. By having responsibility over support staff, observing how colleagues handle management responsibility and studying business and managerial subjects as part of lifelong learning, surveyors can hopefully excel once the opportunity arises. People management and leadership skills are important, and managers need to be able to get the best out of their staff and also any external advisers. Managers will also need to understand economic and market trends, technical issues, business development and marketing issues.

Lateral awareness — case performance

The examples also show the business skill of lateral awareness of the factors affecting a client's interest, but which the surveyor may not necessarily possess a detailed understanding of. General practice surveyors need to be alert to building defects, for example, and call in building surveyors and others with the necessary expertise. If a surveyor acting for a developer misses the need to obtain party wall consent, a footpath informally established on the site, the presence of underground cables, a ransom strip craftily created by a former owner etc, this will be costly to a client.

Lateral awareness — business development

A surveyor working for a property consultancy who understands the range of services provided by his colleagues will be better able to sell other services for them than a surveyor lacking the mental checklist of opportunities, and such instincts to generate further business. An example of such "cross-selling" is an agent acquiring a new investment property introducing the property management team for a client. Also, in return, the management team can use the agency team for lettings. However, in practice, cross-selling has its pros and cons, such as an agency department's reluctance to cross-sell to the valuation department because poor performance could result in lost business for the agency department, and because extensive valuation work may see a client conscious of certain Red Book provisions and the Carsberg Report and no longer wish to employ the agency department. Also, in winning business for other departments, this may enhance their relative profitability, thus affecting bonuses (unless overcome by effective internal arrangements reflecting cross-selling).

Selling skills

When working with clients, opportunities should be subtly taken to reinforce to the client the superior, client-specific, knowledge possessed against potential competitors. Similarly, when seeking new clients and/or pitching for new work, the ability to show an understanding of the client's issues and objectives can enable a prospective adviser to come across well. Surveyors who are able to delight their clients with the level of service available can reap many years of continued business — sometimes being viewed by clients as an integral part of their teams, despite technically being external.

Management Consultancy Roles of Surveyors

As indicated at the start of chapter 3, the management consultancy work of surveyors is where surveyors advise businesses on the strategic management and utilisation of their property assets, having specific regard to the operational nature of their business. Other terms used in the market include "corporate estate management" and "strategic property consultancy".

This chapter initially examines the work of RICS Management Consultancy faculty and then summarises the property issues facing the corporate sector with reference to the "Bootle Report". Illustrations of surveyors' management consultancy work are provided, beginning with a section on sale and leaseback transactions, which shows how management consultancy/business issues and conventional property aspects play a part in surveyors' case work.

RICS Management Consultancy faculty

RICS Management Consultancy faculty states that its role is to "support members involved in management consultancy — ie business solutions to property problems, business management and practice, strategic advice and corporate and personal insolvency and turnaround management". The faculty mentions that "there is no recognised industry definition and, therefore, members and clients have different perceptions of what management consultancy is".

The mission statement for the faculty is:

To establish the members of RICS as the natural strategists within any organisation where property plays a significant part in the business, to support and develop members in achieving the confidence and knowledge to expand beyond their traditional surveying boundaries, to develop and promote best practice in business management, and to support and develop members working in corporate recovery.

RICS Management Consultancy faculty commissioned a research report from the Advanced Management Research Centre at Cranfield School of Management, which examined market attitudes towards management consultancy and surveyors operating in this field, as well as the perceived value placed on property assets. RICS summary publication, *Changing Times — Strategic Consulting for Professional Effectiveness*, summarises the issues as follows:

> Every organisation has three main items of expenditure — staff, information technology and real estate. Naturally, given the tumultuous changes that have rocked the corporate environment over the last 30 years, each of these have, to varying degrees, come under managerial scrutiny.
>
> For example, in the 1980s, organisations came to realise that the way in which people are managed and developed had a huge impact on performance. The Strategic HRM movement led organisations to use high performance HR practices as a source of differentiation and competitive advantage.
>
> Next came information technology. Modern information and communications technology (ICT) has brought infinite possibilities of producing goods and services faster, better and cheaper. New ways of processing information and communicating have fuelled a host of organisational trends designed to release value and encourage efficiency: flatter organisations, outsourcing, globalisation, multi-functional teams, strategic mergers and alliances, and the effective management of knowledge.
>
> The result of such convulsions in the business environment is that few organisations today need convincing that people and information technology are critical to delivering and improving products and services.
>
> However, the third key area of expenditure — property — is the odd one out in this story. Corporate real estate portfolios have expanded rapidly during the last 20 years. Yet the consideration of property as a strategic resource to be managed in line with business strategy has received little senior management attention.
>
> This may be an issue of perception. Property is often viewed as an inherently inflexible commitment, an unfortunate but necessary expense that, unlike people or computer software, is by its nature fixed and unalterable. The idea of using property as a source of added value for organisations remains, at least outside a coterie of specialists, audacious and innovative. Veale, in 1988, suggested only 20 per cent of organisations seek to manage their real estate for a profit.
>
> Yet since the mid-1990s, the first stirrings of a further phase of managerial revolution have begun to emerge — a phase which will witness the potential value attached to the strategic management of corporate real estate becoming increasingly important. Some surveying firms confirm that their roles are starting to blend into those of management consultants. Real estate management consultancy divisions are beginning to flourish both among the large surveying and management-consulting firms. A handful of surveyors are already positioning themselves principally as management consultants with specialist capabilities. Among the pioneers there is a sense of the function diversifying into a far broader conception of chartered surveying than has traditionally been the case. 'Every organisation that utilises working space is also in the real estate business too,' as one survey published in 2000 put it (Corporate Real Estate Practices 2000 annual survey, conducted by Reading University and Johnson Controls).
>
> Outside the world of surveying, too, there is a gradual realisation that property represents a rich source of untapped value for businesses. In a speech to the CBI National Conference in November in 2001, Roger Bootle, the economist, argued that boardroom ignorance of property issues represented a "blind spot" among many senior managers.
>
> The contention of this report is that property needs to move on to the centre ground of business thinking. Just like human capital and information technology, the strategic

management of corporate real estate can be a source of significant competitive advantage for organisations. The fact that it is not yet seen in this light points to a squandered opportunity. In the future, organisations will come to understand that how they manage their property portfolio is just as significant a factor in corporate success as how they motivate their staff.

The RICS publication also provides further illustration on what is meant by "strategic" and comments as follows:

In management conversation, business literature and public presentations, the word "strategic" often tends to be used as a kind of generic seasoning to be thrown around without thought for what it really means. It is worth being precise. What does the strategic management of property mean?

To answer this, it is useful to call on an illustration from the real world that demonstrates the potential of the smart management of property to deliver value — not, in this case, to businesses but to taxpayers.

Local authority assets are valued at around £78 billion, costing approximately £5 billion a year to operate and manage. They include over 21,000 schools, 3,800 libraries, 1,800 leisure facilities and 2,200 community centres (Audit Commission, 2000). Many local authorities are struggling to manage their property portfolios effectively by minimising the cost of property while maximising its contribution to core services. Value for money yielded by these assets is rarely proven, while less than half of authorities regularly review the costs and benefits of their existing portfolio.

The government is already aware of this situation. The Asset Management Planning (AMP) programme is forcing authorities to re-think their property portfolio and how it can be used to better deliver services — a rudimentary version of strategic property management. The AMP process demands that certain strategic questions about property assets be addressed.

It urges authorities to improve their services by ensuring all property is fully incorporated in planning services, encouraging them to review the need to retain holdings if they do not contribute to service objectives. It asks all authorities to capture the full costs of property ownership and occupation, and to dispose of assets that do not justify their expense. It wants them to agree more flexible arrangements for owning and using office accommodation. And finally, it demands all local authorities subject their property services departments to rigorous best value review.

The AMP process is a kind of proto-version of the strategic management of real estate. It seeks to align an organisation's property portfolio with corporate strategy, evaluating how holdings are used against the present and future needs of an organisation.

So, again, what should the strategic management of real estate encompass? In a nutshell, the following: considerations relating to facility location, procurement, the design, construction and maintenance of buildings, physical space planning and layout and all contractual obligations which arise. In addition, it is about the evaluation of strategic need for property, facilities management and end-of-life issues associated with real assets. The real estate function should aim to enable core business, optimise real estate resources; and ensure property objectives contribute to the overall strategy of the organisation.

In short, property represents a colossal expense for organisations. It is a fundamental management issue that they should get value from their investments. Ensuring they receive that value is the function of the strategic surveyor.

The opening line from the RICS publication: "every organisation has three main items of expenditure — staff, information technology and real estate" really means the cost base of larger "organisations" to which RICS management consultancy activities typically apply, and excludes costs such as purchases of raw materials, stocks etc. Also, RICS management consultancy material aims to promote the role of the surveyor, and alert the corporate sector, in the best way possible, to the opportunities often being lost, and the financial gains achievable, from property.

Most surveyors will be dealing with a range of businesses, each with its own market, performance drivers and internal economics. Some businesses will have minimal property costs, and property will be an insignificant proportion of total costs and profit margins. Even then, it may still play an important role in terms of location and the image projected to its clients and staff — and therefore influence profitability. A large manufacturer may be more concerned about the strength of the pound, the consequent competitiveness/price of exports, and the availability and cost of skilled labour than marginal savings in property costs — but, again, property decisions may still make a difference.

By way of further illustration of surveyors' need for new business skills, the RICS publication includes the following information, under the heading of "Earning and learning".

The management of infrastructure has serious implications for both individuals and organisations in terms of employee salaries and motivation, productivity levels and, of course, the way information and knowledge is generated and how it flows through an organisation. Yet in order for surveyors to fulfil a more strategic remit, and win the professional fees on offer, a change in the perception of the client-surveyor relationship is necessary. Surveyors will require a thorough understanding of the client's operating environment, their strategic dilemmas and aspirations, their values and the organisational culture the client is attempting to foster. Surveyors will have to be able to diagnose and solve complex organisational and managerial problems in a way that many will not be used to. The solutions they advocate will have to be co-designed totally with the client's corporate strategic aims. They will have to bring innovative management services and be able to apply broad creative knowledge. They will have to balance traditional technical skills with the broader strategic perspective of the management consultant.

A few may be left behind by the revolution. It is evitable that chartered surveyors will not go unchallenged in staking their claim to this new, under-explored market. Competition will come from accountants, engineers, lawyers and property strategists, all of whom will be vying for position. The question, therefore, is how surveyors can place themselves in the vanguard of property-related management consultancy? How can they learn to create client-centred consultation, align real estate with corporate strategy, and manage the consultation process?

At the individual level, firm level and at the level of RICS, professional development is going to have to encompass a new range of inter-personal, relationship-building and management skills. Surveyors are going to need competencies in change management, organisational learning, business operations management, strategy formulation, team-working, communication, listening, problem-solving and evaluation. Such "skills-set" may sound imprecise yet challenging. Yet there is a risk that the unfamiliarity of such "soft" skills may lead some to conclude they can pick up what they need as they go along, feeling their way in the time-honoured traditions of hit and miss. This would be a profound error. Failure to get the knowledge part right — to think through the learning and development needs that will enable the profession to reposition itself authoritatively in the new market — would be reckless, irresponsible and not least a waste of valuable time. Surveyors, like many other professions, need to acclimatise themselves to the culture of continuing professional development. Where they are lacking, and where their professional body needs to concentrate its emphasis, is on how to furnish the learning — the strategic skills — that will enable them to flourish in changed times.

There is another risk, here, too. The move to strategic corporate real estate consultancy could lead surveyors into conflicts of interest: after all, didn't adding value to accountancy services through management consultancy allegedly help fuel the Enron scandal? It is wise to be ready to think through the tensions which may emerge between strategic consulting and more traditional property-brokering, the conflict, in effect, between saving money and adding value. Clients will expect and deserve clarity in the services surveyors provide.

Such warnings, important as they are, should in no way dilute the fundamental message however. It is only by re-focussing themselves on the needs of the new market in strategic property-related management consultancy that surveyors will enable themselves to reap its rewards. And make no mistake, the rewards are there — neither surveyors nor clients need much convincing. For those willing to upgrade their skills and for organisations looking to profit from their real estate, the outlook has rarely seemed more auspicious.

Further details on the training and education initiatives of RICS Management Consultancy faculty are included in chapter 10.

The Bootle Report

The "Bootle Report" was a study commissioned over 2001/02 by RICS from Capital Economics, led by Roger Bootle. The findings, published as *Property in Business — a Waste of Space?*, highlighted how businesses often fail to get the best out of their property assets. The executive summary from the report is as follows:

The wasted billions
This study discusses how UK businesses could save up to £18bn a year through more efficient use of their property, boosting gross trading profits by up to 13% — and looks at the economic benefits of so doing. A further £2bn could be saved with some changes in government taxation policy. The savings would be:

- Between £7.2bn and £9.5bn if owner-occupiers used space as efficiently as tenants
- Up to £6.5bn if all office-based businesses implemented new working practices such as hot-desking
- £1.3bn from more efficient facilities management
- A possible £300m extra from appealing against business rates assessments
- On top of this, if the government were to alter its policy on stamp duty and capital gains tax, business could save a further £2bn.

Property in the economy
The study focuses on the ownership and use of office, retail and industrial space worth about £400bn at market value.

- In 1998, property related services contributed about 9% of GDP — £78bn.
- Construction contributed about 1.3% of GDP, about £11bn.
- Money spent on facilities management contributed a further £13bn or 1.5% of GDP.
- Overall this was a contribution of just under 12% of GDP, just over £100bn.
- This year, the contribution of commercial property will be just over £120bn, 12% of GDP, given the most recent Treasury forecast.

For the business sector, commercial and industrial property is the most important fixed asset.

Boardroom blindness
Property costs are typically high enough to be the second biggest cost, after salaries, facing business. Given this, it is surprising that many businesses do not have an accurate assessment of the property costs they face and that management of property is usually handled at an operational, rather than strategic level.

Tenants tend to monitor their property costs far more than owner-occupiers, who rarely impute rental costs and use their space less efficiently. But total property costs also include the tax costs of property and facilities management costs. These are often poorly monitored and managed both by tenants and owner-occupiers.

Improving data and analysis
The first obstacle to achieving savings is the lack of information that business has about its property. Often the company does not even have an accurate asset register, or current market value, of its property. The study shows the benefits of collecting property information in a systematic way and analysing that information in the language of the Chief Financial Officer — for instance, by looking at costs per business unit and per employee. This means that the performance of property can be related to the business strategy as a whole.

Cost cutting tools
With accurate and comprehensive data established, corporate property managers can use tools such as affordability ratios, benchmarking schemes, measuring different types of property costs and occupational density, and in-house research. Internal rent-charging is a particularly useful but rarely used tool to make managers aware of the property costs they incur, and to encourage individual business units to take responsibility for property costs.

New working practices
Corporate property managers could encourage more efficient use of space to cut property costs. A key example of this is implementing new working practices such as hot-desking and hotelling. These can cut the costs of property usage by reducing the number of workstations a company needs and by reducing the amount of space each employee needs. The gains to be had from implementing these new working practices in the office sector could amount to nearly £6.5bn. In addition there may be very large productivity gains, as the workforce could use the latest IT tools.

Owners should emulate tenants
Ownership structure also has an impact on the efficiency of property usage. Owner-occupied office space is less densely occupied than leased space in every category of office building. In the UK, roughly two thirds of business property is owner-occupied

with the remainder being leased. If all owner-occupiers were to occupy space at the same occupational density as tenants, the business sector would save up to £9.5bn a year.

Possible reasons for owner-occupiers' less efficient use of the property include the illusion that the property is 'free', because there is no rent to remind them that it is not; or an expectation, more appropriate to the high inflation era of the past, that capital value increases will more than compensate for a lavish space allocation.

To own or lease?
Owner-occupiers could achieve more efficient use of space in one of two ways. First, they could choose to lease rather than own their property, taking advantage of the increasingly sophisticated financial solutions for property ownership that allow businesses to structure their corporate assets more flexibly. Sale and leaseback transactions take property off the balance sheet and release cash to invest or reduce debt. While the eventual property costs are likely to be higher, the difference should be outweighed by more profitable use of the money released.

Alternatively, they could remain owner-occupiers, but take steps to become better informed about the property costs they face and actively manage their property usage in the same way that many corporate tenants do; keeping complete records, charging the business a full market rent and regularly checking costs and space usage against published benchmarks.

Long leases
Although tenants utilise property more effectively than owners, and do monitor its performance, the design of the institutional lease is a distinct disadvantage for occupiers who seek greater flexibility. The institutional lease, designed to protect investors in an era of high inflation, fails to respond to the needs of many tenants in today's low inflation but fast-moving environment. Long lease lengths, coupled with upward-only rent reviews, can become onerous on businesses in a downturn. The absence of break clauses further limits the flexibility available to a firm. The deficiencies of the UK institutional lease are particularly striking when compared to standard lease clauses in other EU countries.

If all lease terms were shortened, there would be more potential tenants "circulating" which should allay landlords' traditional fear of not finding a tenant when a lease ends and the existing tenant does not renew.

Tackling the tax burden
The overall tax burden on UK commercial property is £15bn — 5% of the total tax take in the UK and a higher proportion of GDP than in each of the UK's top five trading partners, threatening Britain's attractiveness as a business location. The government should avoid complacency based on the UK's relatively low overall business tax position today, as business taxes, including property taxes, are set to fall in mainland Europe.

The study also shows that businesses could benefit from seeking advice in order to minimise their tax payments; for instance, by appealing against business rates assessments.

Making property more "liquid"
"UK plc" would benefit overall if distortions and frictions in the market caused by the design of government policy were reduced. The recent rises in stamp duty, in particular, have adversely affected the liquidity of the commercial property investment market. By abolishing stamp duty on commercial property, or at least reversing the increases in the

last Parliament, government would do much to increase the liquidity of the commercial property market, at a loss of only £700m a year in revenue.

Again, there is perhaps a slight promotional angle, with some headline points needing more scrutiny. Lease terms, access and security issues, flexibility for expansion and many similar points all need to be considered by surveyors. Although, in theory, businesses can gain from hot-desking and hotelling-type initiatives, people like to have their regular/secure workspace. This is similar to the potential for home-working not being as great in practice as it appears in theory, with people needing people, and a team ethos often being needed in a business. As mentioned in chapter 2, it is always helpful to think about how markets perform and their likely direction: so while there may be savings achievable for individual businesses, if aggregate space was rationalised/property voids created to the level thought achievable, with new occupiers not being found, property values would fall considerably, also impacting on investment markets, new development etc.

The *Changing Times* and Bootle reports are available free of charge from RICS Management Consultancy faculty.

CBI: *Property for business*

In early 2005, the CBI published *Property for business — an essential guide for senior executives*, which is available free of charge from *www.cbi.uk/property*. This guide develops the issues raised in the Bootle Report, and includes sections on defining your property strategy, delivering your property strategy, maximising workplace effectiveness, optimising your property costs, unlocking asset value and regulatory influences.

The key issues checklist from *Property for business* is as follows:

1. Do you have a property strategy that reflects the operational business plan? — A property strategy must be business led. If no strategy exists, or if it does not match the operational business plan, property decisions will not meet the requirements of the business.
2. Can your strategy be effectively implemented? — To ensure your strategy performs, it should be converted into policies and plans which set out what is to be delivered and how. As ever, access to the right skills and resources will be key.
3. Do you have an accurate database of your property assets? — Poor records could lead to properties being "lost" or to costly surprises such as an unexpected lease expiry or rent review.
4. Does the board receive regular reports on key performance indicators for its property? — Good information is key to efficient management. To optimise business performance, a strategic awareness of property costs and future opportunities is essential.
5. When did you last review your property assets to ensure optimum performance? — Property is expensive: it is important that you make best use of it. With regular reviews, you are more likely to be better at managing costs, controlling stock levels and increasing profitability.
6. Are there up-to-date valuations of your property interests? — Companies must

know what their properties are worth. Undervalued property assets flatter returns and lead to missed opportunities to make windfall gains on timely sales.

7. Have you considered how property could be used to improve workplace effectiveness? — People are your most important asset. But patterns of living and working are changing, and innovative workplace solutions can offer significant productively paybacks through improved staff performance.

8. How costly and how difficult would it be to reduce your property portfolio? — Companies with shorter unexpired lease commitments are better able to adapt to changing circumstances. Knowing your weighted unexpired lease commitment — the average time remaining on your leases weighted by the yearly lease rentals — will help you know just how flexible you can be.

9. What is the cost of putting properties you rent back into their original condition? — Lease termination and exit costs are often higher than planned as many contracts require occupiers to restore any deterioration or changes made to the property when they leave. With an active property strategy, action can be taken to significantly reduce these liabilities.

10. Are you aware of the range of regulatory influences affecting your property holdings? — Changes to accounting rules, stamp duty, corporate governance and environmental regulations have moved property up the business agenda in recent years. It is important that you are aware of any statutory and legal obligations relating to your properties — as well as likely future developments.

As indicated in the preface, comment on purely property matters is kept to a minimum in *Business Skills for General Practice Surveyors*. For further detail on property issues, reference can be made to a separate EG Books publication, *Maximise Your Property Assets* (see *www.propertybooks.co.uk*). Sections include property as a business asset, running the portfolio, managing the properties, extracting cash from the portfolio, disposals and acquisitions, with case studies and statistics also being provided. Bulletins produced by the international property companies also provide further information (accessed via their websites, usually under "publications" or "research") and includes market trends as well as research statistics.

Sale and leaseback

As indicated at the start of the chapter, sale and leaseback transactions are a good illustration of how management consultancy/business issues and conventional property aspects play a part in surveyors' case work.

A sale and leaseback involves a company, as owner-occupier, selling a freehold property, and remaining in occupation by way of a lease.

The sale releases capital for the company, which could facilitate investment in the business, reduce the need for bank or other finance, fund property refurbishment/expansion, simply improve the cash position/working capital, or, for example, enable a stock market-quoted company to buy back its own shares.

Whether sale and leaseback is beneficial depends, very specifically, on the factors relevant to the individual company concerned — including operational business issues facing the owner-occupier, financial and accounting aspects and the nature of the property. Property market conditions — particularly the strength of the property investment market — are also relevant. Sale and leasebacks are more commonly undertaken by larger companies (or the "corporate sector") with examples including a multinational corporate in respect of its manufacturing/ distribution premises, and banks in respect of their retailing outlets. However, sale and leaseback also presents opportunities to smaller businesses owning freehold property, especially when the interest from private investors is strong, as has been experienced over recent years.

Review of property holdings

The surveyor will have examined the company's overall property holdings and determined which freehold properties need to be retained for some time for occupational purposes — and therefore may be suitable for sale and leaseback. The rationalisation of freehold property holdings through vacant possession sale or of leasehold properties through vacation at lease expiry or break, assignment, subletting, or surrender (of whole or part) would not, of course, involve sale and leaseback.

Evaluating the financial gains

Consideration has to be given to the value of the property/ies and background circumstances relating to acquisition/s. If, in simple terms, a property was purchased 20 years ago, without finance/borrowing, at £500,000 and could be sold for £2m today, then £2m would be realised. In contrast, a property purchased two years ago for £1.8m, against which £1.3m is financed, and could be sold for £2m today, would realise £700,000 (ignoring acquisition and disposal costs and other factors). £1.5m profit would be realised in the first example, and £200,000 in the second example. Enhanced reported earnings/profits in a year could be particularly beneficial for a company. Consideration also needs to be given to capital gains tax/corporation tax liability arising from the sale, as this reduces net proceeds (when tax is later due and not on the actual sale).

The financial and accounting benefits of sale and leaseback will also relate to the way that the company has carried the property in its company accounts. Illustrations are apparent from the issues covered in chapter 7, including the way property is treated in company accounts (at historical cost or market value), how the sale/profits of the property are accounted for, and how property can influence balance sheet strength/gearing.

Investment and vacant possession values

A key aspect of sale and leaseback is the scope to sell the property as an investment interest for a greater amount than would be achieved with a vacant possession interest. If, in simple terms, the value of the property with vacant possession is £1.5m, and an investment sale is £2m, sale and leaseback generates an additional

£0.5m. A sale for £2m, and effective utilisation of the proceeds (albeit now paying rent for, say, 15 years) could be preferable to retaining the freehold for 15 years and then selling at £1.5m (subject to changes in value over the period).

Also, the 0.5m difference between £2m investment value and £1.5m vacant possession value would not have been in the company's balance sheet even if propeties had been revalued (which would be on the basis of vacant possession/ existing use value) thus realising greater profit, as well as cash, than the company may have envisaged.

In view of the above, businesses looking for space could acquire a property as a vacant freehold with a view to subsequently securing a sale and leaseback. Indeed, sale and leaseback presents a financing option when acquiring/taking over companies.

Maximising the sale price

The sale price is influenced by all the usual factors affecting the value of investment property. The financial standing/covenant strength of the tenant and the length of the lease/unexpired term/absence of breaks are vital in terms of income security. The type of property, its location and specification, rental growth prospects and reletting prospects at lease expiry are also relevant.

The good quality tenant uses its covenant strength, and likewise its ability to offer a long lease of, say, 15 years to secure a low yield and therefore high capital value. A simple investment valuation may, for example, be a market rent of £140,000 pa, Years' Purchase in perpetuity at a yield of 7% (14.3 multiplier) = £2m. In contrast, if a poor quality tenant occupying an old industrial building commands a yield of 12–15%, this would be an expensive way to raise capital/ finance, and typically be unsuitable for sale and leaseback.

As with the merits of freehold against leasehold tenure generally, the sale denies the owner of any gain in capital value over time, although also prevents any losses. A good time to sell would be when the occupational market and investment market are at their peak, and strong rents and low yields secure the highest capital value.

Although the scope for more valuable alternative development uses to increase capital returns would usually be lost with the sale, the owner-occupier could retain surplus land, retain ransom strips or impose clawback provisions and/or restrictive covenants.

Rent

The rent payable under the lease can be set by the owner-occupier at a full level in order to achieve the highest sale price. There will be more scope to set a relatively high rent with properties that do not have a wealth of comparable rental evidence from lettings, rent reviews and lease renewals, and do not therefore expose relatively high rents or even over-renting.

Rent reviews bring lower rental increases if the initial rent is high, although upwards-only rent reviews could prolong rents set at artificially high levels.

Stepped rents, fixed uplifts or RPI increases at rent review, and hypothetical (specially drafted) rent review provisions could also support the maximum sale price/investment value. The rent also represents a cost that reduces the company's earnings/profit, although this also reduces corporation tax (with the distinction being less had properties been mortgaged and interest repayments charged to the profit and loss account).

Lease terms

The other lease terms need to balance the scope to maximise investment/sale value, and any occupational flexibility needed. Indeed, the owner-occupier's business plan must justify the proposed lease length. Also, although 15 years has been adopted above, sale and leasebacks of certain larger interests require 25 years. The tenant's occupation of the building is now subject to lease terms, which as well as providing the usual restrictions on use, alterations/improvements, assignment and subletting, for example, could mean ultimate liability for dilapidations. Protection can be secured through appropriate repair provisions and a schedule of dilapidations in the lease, but this is potentially detrimental to the investment value. In the case of key strategic property holdings, companies may wish to retain absolute control and not undertake sale and leaseback. Also, where numerous properties are subject to sale and leaseback, wider strategic evaluation is necessary. Groups of properties could be packaged for sale to the same purchaser.

Wider corporate issues

As a general point, corporate occupiers tend to utilise leasehold space more efficiently than freehold space, with rent as an expense helping provide focus on optimum use (including through internal off-charging of property costs to individual departments).

However, property will not always be significant in relation to a corporate's operational turnover and profitability and, despite property efficiencies appearing to be achievable from a surveyor's perspective, the better business decision may be to carry surplus space in order to support unpredictable operational requirements and possible expansion.

Other factors

A brief summary of other factors is as follows:

- "Lease and leasebacks" are similar to sale and leasebacks, and would involve an occupier assigning a headlease and remaining in occupation by way of a sub-lease.
- Capital allowances (tax benefits) need to be considered — presenting advantages to the purchaser, and likewise the owner-occupier, in helping to achieve a good sale price.

- Stamp duty land tax is not payable on the lease in a sale and leaseback.
- A sale and leaseback could be to a property-owning subsidiary of the company, rather than an open market sale.
- Sale and leasebacks may also be undertaken in connection with surveyors' development work — enabling an owner to remain in occupation for so long under a lease, pending planning permission, land assembly and/or the commencement of construction.

Business Skills for General Practice Surveyors provides only a basic overview. In practice, surveyors will liaise closely with accountants and lawyers.

Management consultancy case examples

In order to provide illustrations of higher-level management consultancy work undertaken by chartered surveyors, and how this draws on business as well as property knowledge, the case examples below have been made available by RICS. These also illustrate the operational issues facing businesses in both the private and public sectors.

- **Relocation and government property issues**
  The *Financial Times* reported on July 15, 2003, "The new Home Office Building will be too small to accommodate all its staff ...". The National Audit Office (NAO) had published a report that day into the PFI contract for the new headquarters for the Home Office. The contract was won by a Construction/FM property developer, Bouygues UK Ltd.

  It is not possible to comment specifically on the Home Office situation but of what steps can be taken by organisations, both public and private, to ensure better press for the property industry.

  This issue was explored by Liz Weaver, executive for the RICS Management Consultancy Faculty, when she spoke to faculty chairman, Paul Winter, CEO of Corpra and Neil McLocklin, Director of Strategy at Corpra, a real estate property strategist. The message from the strategists is clear and some would say obvious. The key to success in any relocation or major property is a three-point plan:

  1. Consult with the business units
  2. Align the property strategy to the business strategy
  3. Manage the change process

  Let's explore each of these areas in more detail and demonstrate how they can be translated into a successful PFI or corporate strategy.

  1. *Consult with business units*
  A property strategist would stress the word "consult" — meaning listening, questioning, challenging, and ultimately understanding. The property professionals have traditionally determined requirements with limited knowledge of what exactly the business side does. "Asking what somebody wants is very different from understanding what they need" says Paul Winter. "Challenging a business need to be in London, given the very strong desire to be near the 'honey

pot' of political influence in the corporate or public sector, is impossible without a real insight into the operation".

Modern management thinkers such as Hamel and Prahalad say in their book *Competing for the future* (1994) that "The potential for extracting value out of the management of inter linkages only becomes visible when unit executives from across the company participate in a horizontal strategy development process" (p289).

They describe a "collective strategy" where they say "line managers from across the company come to recognise the potential value-added of collective action" (p289). They cite Hewlett Packard and General Electric as companies that have recognised the potential value of "boundaryless" organisations.

Paul Winter highlighted the need for a four-stage process to establish the necessary insight:

(a) An employee survey across the entire organisation to develop a grass roots perspective and a whole host of quantitative data.
(b) Semi structured interviews with the leadership as well as a cross section of the organisation in terms of function and management level.
(c) Comparison of 1 and 2 to establish the important relationships.
(d) The development of a business map, which highlights the "real" organisational structure and relationships.

The business map will highlight that some business units need to be with other partners, clients, agencies outside of London, while others need to remain within the capital because of much stronger relationships. It provides an objective view of the entire organisation, rather than the biased view of any particular manager.

"Departments or business units can be seen as living in ponds with different eco-systems" says Neil McLocklin. "Whilst the department may be part of a particular fly species (the company or government function), its survival is dependent upon the health of its particular pond and eco-system (its customers, partners or community), not the proximity to other members of its species. Business mapping defines the individual eco-systems".

By creating a business map it shows the flow of information between each employee. An organisation chart doesn't achieve this. What you see in an organisation chart is the reporting structure, vertical. Departments may mix horizontally for flow of information. The type of information a management consultant needs to know is "who talks to who, who do you work with within the organisation?" This exercise often highlights things that managers didn't know! You need to ask "What is the objective of the department". If you look at each department in this way, who feeds who?

The consultation process creates the platform for challenge. Tom Peters, the business guru, often asked the question "How do you ask a customer what he wants when he has not seen it yet?". The same applies to real estate. If an organisation like the Home Office is asked "What should your new building be like?", the answer will be "It needs to accommodate 5000 people in Victoria" because the department has not experienced working in any other way. The next question will be "how will you work in this new building?", and the answer will be constrained by the existing mindset and inertia. The National Audit Office cannot criticise the Home Office Project because it was procured in an appropriate way and merely responded to the brief. It is the initial strategy and brief that needed to be questioned. An Old Organisation in a new building = An Expensive Old

Organisation. The consultative briefing stage needed to address the People; the Organisation and its Workstyle, not just the bricks and mortar.

2. *Aligning the property strategy to the business strategy*
   Once the business map and business strategy is developed and agreed, built upon a foundation stone of participative consultation, then it is important to align a property strategy to it. Clearly each organisation will have a different map and a different strategy, and so again it is difficult to be specific. However, one common thread is often the need for flexibility. This would be highlighted through the consultative process and would appear to be what the National Audit Office criticised the Home Office for. The property strategy was essentially locking the organisation into a commitment to a particular building for many years, when a property strategist would have identified very early on the need for flexibility. The result is a mis-aligned strategy, and the embarrassment of a building that is far too small even before it is half built.

3. *Managing the change process*
   Every property strategy should be considered a change management project by the organisation. How the building is delivered should be simply a project management exercise for example, ensuring it is within the time, cost and quality requirement. The real opportunity is breaking the mould of "Old Organisation in a new building = An Expensive Old Organisation". How do you use this opportunity to create the new organisation?

   To quote Gary Hamel and C.K. Prahalad again "... without a point of view about corporate direction, bureaucracy is likely to be little more than an enforcer of corporate orthodoxies" (p131). "Often the combination of directional ambiguity and tactical orthodoxy poses a substantial threat to future prosperity: ... we don't know where we're going, but we're not going to stray from familiar paths". (p131).

   The success of any strategy is dependent upon encouraging a positive change experience and creating and communicating a clear strategy for change by addressing the following questions:

   Q. What is the story?
   A. A common understanding and truly shared vision of the change.
   Q. Where are we going and how are we going to get there?
   A. Map the change journey, identify the likely obstacles and mood swings and proactively manage them.
   Q. Who needs to come with us?
   A. Identify and engage stakeholders in a timely, considerate, imaginative but credible manner and motivate key personnel to become champions for the changes.
   Q. What will it be like when we get there and will it be worth it?
   A. Model the desired changes in behaviour that will support the change/ transformation.

   A change management plan has a number of key stages. Most importantly, the strategy must be communicated to all stakeholders and the vision of change must be fully understood and accepted by them before any change is implemented. Involvement of stakeholders is the most powerful means of gaining a real understanding of the change from a business and individual perspective.

Conclusion

The property industry is very familiar with the old adage "the most important part of the design is the brief". Strategic consultancy takes the briefing stage back even further into the business to listen, question, challenge and understand the real business drivers. The property industry and PFI contractors cannot be criticised if the brief is not aligned with the business strategy. As organisations become more susceptible to change, like the Home Office, it is critical to emphasis the strategic consultative process before embarking on any property project, especially as the process itself might negate the need for the property project at all!

- **Compliance issues**

A large national support-services company had acquired businesses and various property over several years across the UK. Some of these properties were contract-related, and others part of their general portfolio, mostly commercial offices, but some workshops, warehouses, and miscellaneous property.

It was necessary to establish whether the various businesses and business managers were complying with key property-related legislation and to what degree was compliance being achieved. A software system was developed to form part of an audit process for legal compliance, incorporating a detailed scoring system and benchmarking process across ten specific areas of compliance. The objective was to establish a "Strategy for Compliance" across the Company, and to compare properties and locations, to track future improvements and to project "what-if" scenarios via simple computer models.

Ten sample, but significant UK properties were initially audited, partly by interview of managers and partly by obtaining evidence to almost 100 key questions on site. Following the property audit process, reports were generated, amplified with graphical charts, scores, as well as detailed text. It was immediately apparent that whilst some legislation was being complied with effectively, the Company was exposed to varying degrees of risk in key areas of non-compliance.

For the first time managers could see which areas had to improve in order to reduce legal (and other) risk, where to apportion budget-spend as priorities, and could set targets for year-on-year improvement as part of the benchmarking review process.

Whilst there was some initial negative reaction from facilities managers who saw the results as a perceived criticism of their management, over time they accepted there were clear benefits in being able to work towards a "Strategy for Compliance" with clear objectives and a common approach. Indeed, they were able to use the independent reports as evidence that some areas beyond their control within the organisation needed urgent change and improvement. For example, there was no access policy on the properties, so there was no incentive to comply with the legislation, or carry out investigative audits.

One facilities manager was given by her director, as part of her annual appraisal, objectives and targets from the previous audit of her premises as a goal to achieve, and exceeded all expectations in putting the appropriate systems in place to reducing exposure to legal risk as a continuing improvement over time.

The audit reports, when read in conjunction with other information, aided strategic decision-making at the most senior management level, yet provided useful practical advice at office manager/FM level.

The consultants who produced the property audit system are now established as a management consultancy practice in strategic property advice including accessibility, property auditing, and other legislation-compliant consultancy.

- **Corporate survival**

  At the beginning of 1997, as a result of rapid expansion by takeover and consolidation a FTSE 250 networking and software systems company, making trading profits of £6.9m per annum, had 75 office, industrial and warehouse properties around the UK totalling 1 million ft² (92,903 m²). 34 properties were surplus and empty costing over £8m per annum in rent and empty rates before ongoing maintenance and dilapidation liabilities; and efforts by existing agents were making no headway.

  In competition with one of the world's largest real estate agents, a boutique firm of chartered surveyor management consultants were appointed to formulate a rationalisation strategy, advise the board and implement the exit from 510,000 ft² (47,436 m²) of property. After inspecting, analysing and subjecting every property to the management consultant's in-house cashflow model in accordance with FRS12 on the consultant's advice, supported by auditors Andersen, the client took a write-down of £32.2m against surplus property. The chartered surveyor management consultants then appointed and managed over 30 firms of agents around the UK to effect the disposal of the individual properties in line with a clear strategy for each. This involved a mixture of methods including lease restructuring, obtaining change of use or new planning consent, purchase from the landlords and back to back onward sale, lease surrender, lease assignment, sub-letting of whole or part, relocation, freehold sale, investment sales of over £20m, refurbishment, redevelopment and re-letting.

  Within 24 months the team led by the management consultants disposed of 51 buildings totalling 726,302 ft² (67,476 m²), 90% of the surplus, producing revenue savings of over £10.63m pa. They also identified a number of development opportunities, including an 8.4 acres donkey paddock at Egham now sold to a major housebuilder. In another case the client made a profit of £12.5m over book value.

- **Production oversupply, wrong location and incorrect facilities**

  The customer is involved in an industry sector where there is strong global growth but a decline in the UK based industry. This is because of the difficulties of achieving economies of scale in the UK and also the high cost base. This company has an oversupply of production capacity in the wrong locations, facilities of the wrong size and aged and functional obsolescence. The original capital cost of these production facilities was very high but the fabric of them is dilapidated and a significant proportion of the facilities are approaching the end of their economic life. The company identified a need to shift production over a period of time but to retain the current facilities in the short term to meet operational needs.

  Another key issue for the company was to capture the full value of the sites — both current value and upside potential. A firm of chartered surveyor management consultants was called in by the company to help to build a strategy. The key issue was to meet the commercial needs and commercial objectives of the company. The chartered surveyor management consultants investigated the company's operational needs going forward and developed models which quantified these needs. Also modelled was the need for liquidity and capital mobility, underlying asset values and the potential full value upside.

  After considering and modelling a range of scenarios, the chartered surveyor management consultants selected an innovative development partnership and leaseback model. This enabled the customer to realise current underlying asset value, maintain the operational facilities as long as their operation was economic and also to capture full development potential in the future. The operating company could therefore exit from any of their sites at any time according to their own industry circumstances and corporate needs.

- **London authority leisure facility**

  A London local authority had a major leisure facility which had failed under public ownership, then failed under private ownership, and had therefore fallen back into public hands. There had been fatalities at the park and, as a mothballed structure, posed further health and safety threats. Also there was no realistic prospect of a management team being identified to run the park on commercial terms without massive subsidy from the local authority. The local authority believed that the problem was an unsuccessful water flume theme park that needed too much public subsidy and how could it be modified to acquire less subsidy?

  A firm of management consultants employing mainly chartered surveyors trained in management consultancy was asked to consider the problem. They were able to persuade the local authority to reconsider the current and future needs of the local population. They identified that there was an under supply of standard 25 metre, laned, swimming facilities, insufficient gymnasium and aerobics provision for lower income groups, inadequate childcare (crèche) for local working parents, inadequate public transport links and a general lack of modern leisure facilities in the borough.

  This consultancy considered a variety of options that were modelled and tested. The only solutions which were to be considered were those that involved no further funding or any capital investment from the local authority. Apparently a "wicked problem", ie one with no practical solution.

  The solution selected was agreed and then market tested by the consultant to prove to the council that the solution was feasible and worthwhile putting back to the market. This was agreed. The consultants then identified a partner to facilitate this solution and provide finance, technical and professional expertise. The consultants identified that the cost of a new swimming pool, gymnasium, crèche and transport interchange could be matched by the development of an upmarket health and fitness club with swimming facilities, a modern family-themed restaurant with childcare facilities and also made a small profit for the financial partner.

  The property strategy clearly met the objectives of the local authority financially and operationally. However, the project needed to be guided through a public consultation process and satisfactorily delivered. The leisure obligations of the local authority have been met and exceeded without subsidy and all of the objectives have been achieved, whilst still retaining public ownership of the key facilities. Management consultancy surveyors advised on strategy, scenario modelling and testing and implementation of this project.

- **Change management**

  An international company owned its European headquarters site. The company was on the point of substantial change. This was brought about through the introduction of significant e-commerce projects which are expected to impact heavily on the business processes.

  The company's site is in an attractive location and is a campus style. It is partly sub-let to other occupiers. The locality has traditionally had low commercial value pricing and a low level of both occupier and investor/developer demand. The challenge for the chartered surveyor management consultant was to build a strategy which captured a thin level of current demand, improved the underlying asset value and meets the operational needs of the company as they change. The main element of the chartered surveyor management consultancy work was to separate the funding/corporate finance issues from the functional needs of its customer.

The management consultant developed a variety of separate models. These were tested in both traditional and innovative ownership models. The overlaps in the separate strategies pointed to the need for a step change in the nature of the site and the potential capturing of new demand. The innovative ownership development structures which formed the heart of the chosen model is ground breaking. By combining the physical aspects, the services package and the financial and insurance products, has led to a "product"'which stimulates demand backed by a major plc with emerging spin out companies. This has transformed the underlying asset value of the site which is a key facilitator to the improvement in working environment and also in attracting significant business synergies.

- **Developing an acquisition strategy to meet business objectives**
  A globally successful retail franchise business, No.1 in the world in business services, originating from the United States was experiencing problems unique to the UK. The rate of rollout compared with other North American and European countries was much slower than programmed. They had been using traditional estate agency and brokerage suppliers to identify properties but still ailing to meet the business objectives.

The company's management identified a firm of management consultancy chartered surveyors to look at the reasons or this failure and to develop an acquisition strategy which would meet the business objectives. The company identified a number of key structural problems in the roll out process and devised a complete re-engineering of the acquisition process.

The US parent had not taken into account the peculiarities of UK landlord and tenant law and more particularly the precedents and conventions of retail lettings in the UK. The nature of the UK landlord style type and term of leases was identified as an obstruction as well as the legal procedure to conclude these contracts.

Furthermore, because of the franchise nature of the business it was not possible to dislocate the finances of each acquisition from the property transaction and the network of landlord was reacting badly to this.

It was not being particularly well articulated by the company's brokers. The company devised a programme of corporate core acquisition as a means of setting a new standard, building the brand awareness amongst the landlord and brokerage communities and developing an accurate location model to identify towns and locations where profits would be maximised in relation to high or low rents.

Using this process and implementing the mission critical elements themselves, this consultancy established a core corporate retail base for the company, taking the total to over 40 locations and returned the brokerage function back to the UK brokerage market, allowing the franchisees to deal directly with their own regional broker but according to a new process and set of rules.

This property strategy was evolved and examined in cooperation with the business using strategic tools, management consultancy processes and micro-economic modelling techniques.

The Lyon's Review

The Lyon's Review, undertaken in 2004, was an independent study into the scope for relocating a substantial number of public sector activities from London and the South East of England to other parts of the UK. Property is a key element in the opportunities to relocate, and the press notice announcing the publication of the Lyons' Review stated:

Sir Michael Lyons confirms that government departments have identified some 20,000 jobs that could move out of London and the South East and recommends that they should urgently take forward their relocation plans in the 2004 Spending Review. A further 7,000 posts would no longer be required as a result of efficiencies. Potentially, over £2 billion could be saved over 15 years as a result.

He concludes that the pattern of government has to be reshaped. The concentration of national public sector activity in and around London is no longer consistent with Government objectives and does not reflect the large cost disparities between London and the rest of the country or benefits of dispersal for the efficient delivery of government business or for the regional economies.

Sir Michael acknowledges that London, as capital, needs a governmental core supporting ministers and setting the strategic policy framework. However, in every other respect, the status quo is open to challenge. And, if Government wishes to make a significant impact on the pattern of its locations across the country, it will need to take firm action.

Sir Michael's executive summary is set out below and illustrates some of the detailed combinations of business, economic and property issues, together with social, political and other elements that will arise in connection with surveyors' public sector work.

I was asked to advise ministers on the relocation of public servants out of London and the South East.

I conclude that the pattern of government needs to be reshaped. National public sector activity is concentrated in and around London to an extent which is inconsistent with Government objectives.

In particular this pattern fails fully to reflect the large cost disparities between London and other parts of the UK and the revealed benefits of dispersal for the efficient delivery of government business and for regional economies.

London as capital needs a governmental core supporting ministers and setting the strategic policy framework. In every other respect the status quo is open to challenge.

If the Government wishes to make a significant impact on the pattern of its locations it will need to take firm action. I have proposed ten recommendations as follows:

1. Departments have identified more than 27,000 jobs that could be taken out of London and the South East, including up to 20,000 jobs for dispersal as a first tranche. Plans for these dispersals should be taken forward urgently as part of Government's forthcoming spending review.
2. Major dispersals are unlikely to offer a quick payback and they incur considerable costs up front. The Government must be prepared to make the necessary investment. Equally, there is a strong case for sharper incentives to encourage

departments to seek the benefits of locations out of London and to keep their presence in the capital to a necessary minimum.

3. Departments should implement their relocation plans alongside efforts to align their pay with local labour market conditions. My review has demonstrated that failure to make progress on locally flexible pay will limit the efficiency gains from dispersal, and could undermine the economic benefits for receiving locations.

4. Whitehall headquarters should be radically slimmed down, reflecting a clearer understanding of what is really needed in London, and of the distinction between policy and delivery.

5. There should be a strongly enforced presumption against London and South East locations for new government bodies and activities; for functions such as back office work and call centres which do not need to be in London; and for bodies and functions whose effectiveness or authority would stand to be enhanced by a location outside London.

6. Cabinet needs to give continuing political impetus to the locational agenda. Leadership should be provided by a Cabinet Committee and, in the short term at least, a lead minister. These arrangements should be supported by a small, short life unit at the centre, to act as a ginger group, to monitor and report on progress with dispersals, and to ensure that best practice is disseminated and embedded.

7. Permanent secretaries and other public sector chiefs are responsible for managing their departments' resources, accounting to ministers and to Parliament. Locational considerations must be an integral part of these responsibilities. The aim should be to mainstream the locational aspect of business planning.

8. The Government must take responsibility for the whole pattern of its locations, developing a strategic framework of guidance for departments and ensuring a mechanism for reviewing and where necessary challenging departments' locational preferences.

9. The Government office portfolio must be much more tightly managed. In particular, exits from London should be coordinated to ensure overall value for money and to strengthen individual relocation business cases.

10. The civil service needs a more coordinated approach if it is to minimise the costs and the adverse impacts on staff associated with relocation and redundancy.

These actions will help create a better pattern of government. By setting a good example, the Government may also promote more rigorous thinking about location in the wider economy, in the interests of UK competitiveness.

Introduction

1. In government, as in business, location matters. Where government is placed has an important bearing on the value for money it secures for taxpayers, the quality of services it delivers for customers, and the legitimacy it earns in the eyes of citizens. Location has implications for local and regional economies and for the character of the government service. All this is highly relevant to the Government's current ambitions for improved public services, efficiency, regional competitiveness and devolution.

2. The forces of technology and global competition are changing the conditions in which government and citizens interact, and a wider debate is being conducted about inequalities between and within regions, and about the concentration of power in London. In the past, the geography of government has received intermittent bursts of attention, but never been subject to sustained challenge. A different approach is likely to be needed in future.

3.   That is the context in which I was asked, in April 2003, to review the scope for relocating public sector jobs out of London and the South East, and that is why the task is important. I regarded the Chancellor's 2003 Budget statement, which announced my review, as marking a watershed in public policy on locations. My work generated considerable interest and elicited many "bids" from local and regional bodies. I saw this response as evidence of a lively public interest in power and diversity in the UK, and of eagerness from authorities across the country for economic growth and greater influence.

The policy and historical context

4.   The Government is committed to improving the efficient delivery of public services, boosting regional economic growth and bringing government closer to the people, through greater decentralisation and devolution. My review is relevant to each of these themes and I have taken note, in particular, of plans for elected regional assemblies subject to the outcome of public referendums, and the Government's stated ambition to empower local authorities.

5.   There are also significant external factors. Technology will continue to change the character of government and the ways it relates to citizens. The office environment is evolving, with home-working, hot-desking and other kinds of flexible working becoming more common. Global competition has already increased the international mobility of many types of work, and the public sector will clearly not be immune from future changes. Following the events of September 11 2001, the need for resilience in the face of emergencies should have become a more prominent strand of business planning.

6.   What does all this portend for the pattern of government locations? It suggests that future dispersals from London and the South East (and reconfigurations more generally) are likely to be part of bigger reforms which also transform the nature, organisation, productivity and size of public service functions. They may come about by a number of routes, including decentralisation, devolution or a change in the boundary between public and private sectors.

7.   These moves are unlikely to follow the model of earlier government relocation drives — in particular the dispersals spawned by the Hardman review of 1973 which transferred self-standing business units to pre-ordained locations in the interests of regional policy. They helped give rise to a narrow and mechanical conception of "relocation" — a kind of chess game played within the machinery of government.

8.   This approach is outdated, as is the term "relocation" itself. For modern times the "locational dimension to business planning" might be nearer the mark. I see my review as championing good business planning in government and the responsibilities of service chiefs to deliver.

Why disperse? The impact on government business

9.   London remains the most expensive part of the UK for doing business, and an often difficult place to find employees. There is no sign of that changing in future. The financial modelling carried out by my review suggests that relocating 20,000 posts could save the public purse more than £2 billion after 15 years.

10.   The evidence is clear that organisations which have dispersed activities from London and the South East enjoy significant cost savings, reductions in staff turnover and improvements in the quality of service they deliver.

11.   The business case for dispersal is not just about cost savings. New locations can provide the spur for new ways of working: adopting better business practices,

processes and technology, and reforming organisational culture. The best relocations seem to have been pursued as part of a broader reform and re-engineering effort.

12. Dispersal is never problem-free and there are particular issues facing split headquarters functions, for example the amount of senior time spent in visits to London. The research suggests that clear leadership and careful management can contain and reduce these problems (for example by establishing clarity about the real need for meetings, and fully exploring the potential for alternative modes of communication).

13. Careful attention needs to be paid to the impact on individuals. Modern family structures have created a complex context for location decisions. There are increasing numbers of dual income households and workers with caring responsibilities, and people are protective of their work/life balance. But the benefits for individuals should not be dismissed. Staff who move in post may stand to enjoy large improvements in their quality of life and there are benefits for those living outside London and the South East too — new job opportunities and the chance to pursue a public service career without having to move to London.

Why disperse? The impact on communities

14. The economic analysis I commissioned confirms that dispersal of government activity is likely to bring positive economic benefits for receiving locations — a more optimistic prognosis than that of Sir Henry Hardman in 1973.

15. But there are some important conditions. The impact — measured in terms of knock-on job creation — will be greater when dispersals maximise the business benefits to the organization and where they are clustered in a limited number of locations rather than very widely spread. The impact is also greater where pay is aligned with local labour market conditions, so that relocated jobs are not at risk of crowding out or bidding up the cost of local jobs in the public and private sectors. In the absence of local pay flexibility and a degree of clustering in dispersals, the long-term economic effects of government dispersals may be much smaller.

16. There are also wider spin-off benefits associated with bringing new investment, jobs and people to particular areas — for example the potential to regenerate run-down areas, build public sector career hubs and revitalise civic institutions and community action.

17. What is the impact on London and the South East of exporting jobs to other parts of the country? The evidence is that the disadvantages to London are likely to be short-lived and outweighed by the benefits to other areas. Authorities in London have broadly supported this analysis. It has been put to me that some 20,000 jobs taken out of London would be negligible set against a labour market of several million and forecasts of strong economic and employment growth, and that relocating jobs may help to relieve some of the overheating in London.

18. Some parts of the South are deprived and they are not all necessarily suitable exporters of government work. Care must also be taken in relation to the impact of dispersal on ethnic minorities in London. These issues are explored in the main report.

The scope for greater dispersal

19. Starting with a clean slate, the Government would be unlikely to replicate the current distribution of functions across the country. Certainly, there is wide dispersal, and this reflects in part the previous wave of relocations as well as

continuing efforts at dispersal, albeit piecemeal and low-key, up to the present day. But the lack of sustained focus on location as an integral feature of government business planning has ensured that the pattern is not optimal and is still too dominated by the pull of London.

20. Striking features of the current distribution of government activity include very large Whitehall headquarters; a heavy London concentration of senior level posts; a significant residual component of back-office and transactional work, including call centres, in London; and a surprising number of arm's length bodies, regulators and inspectorates still in the capital.

Departments' proposals

21. I asked government departments to submit relocation proposals. About 27,000 posts could be taken out of London and the South East, of which up to 20,000 are candidates for dispersal. Taking account of expected job reductions because of efficiency measures, the net job creation elsewhere in the country would be likely to be a lower number.

22. Departments have made a promising start but now need to convert these proposals into plans. Nor have they exhausted the full scale of the opportunity for dispersing functions out of London and the South East. The 20,000 posts are best viewed as a first tranche. The proposals:

    • Mostly involve the movement of relatively junior, operational posts.
    • Show a preference for departments' existing regional sites.
    • Leave substantial scope for reducing the size of departmental London headquarters, and for reconsidering the London headquarters location of many executive agencies, arm's length bodies, regulators and inspectorates.
    • Leave further scope for relocating back office and government call centre activity out of London.
    • Leave wide scope for dispersal opportunities arising from joining up functions across organistional boundaries.

23. Departments were disappointing in their stance on headquarters and policy functions. With some exceptions, I found that they had not challenged sufficiently rigorously the case for activities to remain in London headquarters. It is not a caricature to summarise some of the responses as "we're here because everyone we deal with is here"; "it's policy, so has to be in London" and "everything we do in London is indivisible".

24. Such thinking reveals the need for greater clarity about the essential constituents of headquarters functions and more precision in defining policy. There is a need to confront outdated and pessimistic views about the potential afforded by communications technologies, including videoconferencing, and to challenge perceptions shaped by a Whitehall mythology about earlier relocations.

Reshaping the pattern of government — an agenda for action

25. I was asked to advise ministers on the relocation of public servants out of London and the South East. It is not my business to prescribe in detail how this should be done. I have recommended a broad approach which is consistent with the evidence and with the Government's own objectives.

26. The geographical pattern of government activity needs to be reshaped. National public sector activity is concentrated in and around London to an extent which is inconsistent with Government objectives. The current pattern fails fully to reflect the

large cost disparities between London and other parts of the UK and the known benefits of dispersal for efficiency and delivery. It does not reflect the Government's regional ambitions or policies for decentralisation and devolution. Nor is it best placed to maximise resilience in the face of terrorist attack.

27. London as capital needs a governmental core supporting ministers and setting the strategic policy framework. This is not an endorsement of the status quo, which in every other respect needs to be challenged.

28. There is an immediate need to convert the departmental proposals to my review into firm plans using the leverage of the coming spending review.

29. The key to progress is firm leadership. Ministers must be actively involved in helping to reshape the pattern of government. Permanent secretaries and other public sector chiefs are responsible for managing their departments' resources, accounting to ministers and to Parliament. Locational planning should be integral to their responsibilities, and reflected in their formal duties, accountabilities and performance management arrangements. The eventual aim is not "relocation" but the embedding of the locational aspect of business planning. This may also need new incentives.

30. In three key areas, Government needs to become more coordinated to get the best outcomes. The overall geographical pattern of locations has a major bearing on the Government's objectives for efficiency, the reform of public services, regional economic growth and devolution. The Government must take responsibility for this evolving pattern, rather than be content for departments to pursue their separate locational plans without reference to each other and to broader concerns.

31. Secondly, the Government office estate must be much more tightly managed. In particular, exits from London should be coordinated to ensure overall value for money and to strengthen individual relocation business cases.

32. Thirdly, the civil service needs a more coordinated approach if it is to minimise the staffing costs and adverse human impacts associated with relocation. It makes business and economic sense to pursue an approach which emphasises moving posts rather than people, and to seek to redeploy staff who do not move with the post, rather than to make them redundant. It also makes sense to build up activities in other locations, as well as relocating existing activities from London and the South East.

33. I have set out ten recommendations at the head of this summary and in more detail in chapter 10 of this report (not reproduced here). If my suggested plan of action is followed, there will be a real prospect, in time, of government becoming better placed to deliver on its objectives.

A full copy of the report is available from *www.hm-treasury.gov.uk/consultations_and_legislation/lyons/final*. Crown copyright is reserved in respect of the above extract.

The Economy

Surveyors' understanding of economic trends helps enable the opportunities and risks in certain property dealings to be evaluated.

The level of interest rates, for example, influences the cost of domestic mortgage finance and therefore house prices. In turn, house prices influence the viability of residential development, the value of residential development interests, and the scope for a property owner to secure increased sale receipts from higher value alternative/residential uses (such as in the case of a secondary office being converted into residential flats).

In the commercial property market, the cost of finance affects investment yields and the capital value of investment property. In turn, investment values affect the viability of commercial property development, the value of development interests and the level of new development taking place. New development, such as offices, adds to the supply of office space in a local market, and can lead to oversupply — thus affecting rents, investment returns and development viability.

The profitability of property consultancies, the strength of the jobs market and the salaries of surveyors are all influenced by the state of the economy. As indicated in chapter 2, surveyors with an entrepreneurial eye, looking to grow investment, development or professional services businesses need to be able to read markets and exploit opportunities that others are less able to see.

Global economic trends increasingly affect the UK economy. Exchange rates, for example, affect the competitiveness of UK exports and UK manufacturing output — in turn affecting demand for industrial space, rents and yields. The performance of the UK financial markets, which are influenced by global markets, affect the demand for office space, rental levels and investment returns in the City of London, where the financial sector is concentrated.

Set out initially in this chapter is a reminder of some of the economic issues that surveyors are likely to have studied at university, together with illustrations of their application in practice. An overview is then provided of the main economic variables and how the economy is managed, with reference also being made to Economic and Monetary Union (EMU). Examples of economic commentaries are then provided in order to illustrate how the various issues covered in the chapter interrelate in practice, and a property market commentary further illustrates its dependence on wider economic issues.

Introductory economics

Surveyors' education in economics usually begins with theoretical university study. While elements can appear unnecessarily academic and not very relevant to market trends such as those mentioned above, there are still principles that can be drawn on in property dealings and in running businesses.

Supply and demand

"Supply and demand" is a phrase often used when talking about markets. The balance between supply and demand influences price levels in markets — such as the market for office space or the market for particular surveying/professional services. As a general rule, on the demand side: an increase in demand leads to higher prices; and a decrease in prices leads to greater demand. On the supply side: an increase in supply leads to lower prices; and an increase in prices leads to greater supply. Exceptions include increased demand resulting from higher prices because the perception of greater quality and/or exclusivity, including through "snob value".

Elasticity

The sensitivity of supply and demand to changes in prices is termed "elasticity". Demand would be price elastic if, for example, a 10% reduction in price led to a 50% increase in demand/sales. An example of "price inelasticity" would be where a 20% price increase led to only a 5% reduction in demand/sales. Supply would be inelastic where supply changes relatively little in relation to price changes, and elastic where supply responds sensitively to a price change.

For property, demand is typically sensitive to rents and sale prices (ie elastic). This is particularly so in markets where consumers/tenants can chose between a number of properties, and where there are mechanisms providing price transparency and knowledge of all other factors to all market participants. If, for example, new lettings are being achieved for the best-quality West End offices in the region of £75 per sq ft (psf), agents/market participants will be well aware of this, as they will be of other available offices in the area, their precise location, specification and other attributes. If a property is overpriced at £80 psf, there is likely to be no demand. If a landlord is keen to secure a tenant, and prices the rent at £70 psf, strong demand should see the property let quickly. If the supply of vacant offices becomes squeezed and tenants do not have other choices, a tenant may have to take on the £80 psf property. Similarly, if there is strong demand and tenants have to compete for the same property, best offers should see the resulting price reach a higher level than if the landlord is negotiating with only one tenant. What all this means to the surveyor is that the asking rent needs to be realistic (extending also to other terms, including tenant incentives). Moreover, the surveyor is considering price elasticity.

As another example, in the housing market homeowners considering selling sometimes believe that their property is worth more than it really is — because they especially like it or simply because of their greed for the highest sale price. The property can stay on the market for months, and only eventually be sold because house prices rise generally, and catch up with the homeowner's original, unrealistic aspirations. There is price elasticity because there are similar homes on the market, and there is price transparency. On other occasions, an agent can reluctantly put the property on the market at a high price insisted by the homeowner, only to be surprised that offers at the asking price are immediately received. This is more likely for one-off properties where value is difficult to accurately assess, as there are no sales of similar properties or comparable valuation evidence.

When running a property consultancy, it is important to gauge the sensitivity of pricing in relation to the number of instructions won. A surveyor/consultancy with a strong track record providing niche services to the private sector may see demand not being particularly sensitive to price, whereas surveyors/consultancies providing relatively standard services to the public sector may win business only if they quote the lowest fee. The niche consultants' alertness to price inelasticity, and particularly the scope to raise fees, could have a considerable effect on profitability.

Competition

"Perfect competition" is an economic concept that includes numerous suppliers in the market, perfect knowledge among market participants (including price transparency) and no barriers of entry to the market (such as entry and exit costs, or other resources). In contrast, a "monopoly" is where there is only one supplier. Whereas suppliers will be "price-takers" in competitive markets, they will be "price-setters" in monopolistic markets.

In practice, suppliers fall between the two extremes — such as the above example regarding consultancy services where the niche operator may enjoy some monopolistic elements, and the supplier of standard services is suffering severe, but not perfect, competition. Naturally, surveyors and businesses seek monopolistic opportunities ahead of unduly competitive markets. Exceptions, however, include the scope to profit considerably in a crowded, but large, market through various competitive advantages, as opposed to the possibly low level of business and profitability achievable in a monopolistic, but small, market.

It is rare for a monopoly to exist in the property industry, as usually there is always competition to some extent. Monopolistic elements can, however, be seen — really meaning that substantial competitive advantage has been secured, extending to the ability to earn "abnormal profits" — instead of "normal profits" generally available to the market at large. An example could be a supermarket acquiring two potential sites in a town, knowing that only one of the two sites will secure planning consent, but competition from another supermarket will be eliminated, and one site guaranteed to win consent. The prevention of competition also helps limit the level of planning gain contributions — ie the additional offerings to the community that can help swing planning decisions in the developer's direction.

In order to uphold monopolistic advantages, "barriers to entry" need to be strong. Pubs, bars, restaurants and coffee shops, for example, can be established in many areas relatively easily, and quickly expose existing operators to competition. However, if planning restrictions are tight, it may be impossible to establish new uses in an area, and existing operators can continue to enjoy favourable trade.

On behalf of their clients, surveyors need to be alert to both competitive threats and defensive advantages, as this will influence the potential of a client tenant's or owner-occupier's business — and therefore the amount they are prepared to pay to secure the opportunity. Such factors tend to be more relevant in the case of licensed and leisure operations, and the retail sector, but for other property types, location, proximity to clients, availability and cost of nearby labour, transport costs saved through accessibility etc can present considerable competitive advantages to businesses.

Costs

As a further example of the surveyor's need to understand the economics of a client's/employer's business, costs need to be considered.

"Fixed costs" remain the same, irrespective of the level of sales/turnover — such as property overheads and a secretary's salary in the case of a small business, or external repairs and insurance in the case of multi-let offices.

"Variable costs" change with sales/turnover — such as the cost of retail goods purchased from suppliers, or the overtime/increased salary paid to staff for additional work.

Businesses consider "total cost", "average cost" and "marginal cost". Average cost is the cost per individual unit, such as the cost of building each home on a housing development. Marginal cost is the cost of increasing production by one unit, such as the cost of building one further house — and which will be a variable cost. A similar example could be a property consultancy taking on a further single instruction.

When fixed costs are relatively high in relation to variable costs (and therefore to marginal costs), it is more likely to be viable to increase production/sales/turnover. A manufacturer may, for example, have sales of £1m, fixed costs of £700,000, variable costs of £200,000 and a profit of £100,000 (10% profit margin: £100,000 ÷ £1m). If sales/production is increased by £100,000, the additional variable costs will be £20,000, producing £80,000 profit on the extra sales/production. This is a profit margin of 80% for the extra production, and lifts the overall profit margin from 10% to just over 16% (£180,000 ÷ £1.1m).

This relationship between sales/income/turnover, fixed costs, variable costs, profit and profit margins is why good and bad years of trading can have a particularly sensitive effect on profitability. In the above example, a 10% increase in turnover leads to an 80% increase in profits. Had turnover fallen by 10% to £900,000, fixed costs would have remained at £700,000, variable costs would have fallen to £180,000 and profit would fall from £100,000 to £20,000. The effect is most sensitive with businesses operating at low profit margins, and/or having a high proportion of fixed costs.

As another example, a property consultancy may, for example, have turnover of £1m, costs of £800,000 and profits of £200,000, producing a profit margin of 20% (£200,000 ÷ £1m). If further instructions are available at £50,000, the business can accommodate this without taking on further surveyors, and the surveyors responsible for the instruction receive a fixed salary without bonus. Variable costs may therefore be negligible, and a 5% addition in turnover of £50,000 raises profits by 25% to £250,000. (Gains achieved through greater productivity/efficiency and cost-control have similar effects.)

In the above examples, an additional unit of revenue, known as marginal revenue, is greater than the cost of the additional unit, known as marginal cost, and increased production/sales is viable. This shows how "economies of scale" work — with average costs decreasing as sales increase and profits consequently rising.

The examples also show how a business will have a certain level of sales at which it breaks even (ie £nil profit), but beyond which profits will come through strongly. As indicated, this will be particularly so for businesses with a higher proportion of fixed costs. Although capital/investment costs may be seen as an obstacle to establishing a business initially, or developing an existing business further, this acts as a barrier to entry to others, and is therefore defensive against potential competition. Other defensive qualities include brand names and the expertise of key employees. A further illustration within the above examples is that the high profit margins available from additional production/sales provide scope to reduce prices, which for goods/services that are price elastic will help sales and profits expand considerably.

In contrast to the advantages of economies of scale, "diseconomies of scale" involve an increase in marginal costs, and therefore average costs, as sales increase. This reaches a point where marginal cost exceeds marginal revenue, and further sales are loss-making and not viable. As an example of the many interrelated economic effects, greater supply may also reduce price/revenue — not only marginal price/revenue, but also the price of all goods, thus affecting average price and total revenue. Prices can therefore be falling at the same time as costs are increasing, having a twofold effect on profits. Also, in the same way that supply and demand affect price and revenue, it will also affect costs such as labour and materials.

In order to facilitate the expansion of a business, property may need to be extended or relocation may be required. The surveyor will need to work closely with the client business to assess the opportunities and risks, much of which will relate to the market in which the business operates and to its internal economics. The business will wish to see how property costs fit in with all of the business' other costs, including consideration of whether these costs are primarily fixed costs or whether there are ways for property costs to be as variable/flexible as possible. Flexible leasing arrangements may help, as opposed to a freehold acquisition, which may not be feasible to let out to tenants in part when not used to full capacity. As indicated in chapter 3, and as an illustration of the need to examine the components of total, average and variable costs, property issues are most crucial where property represents a larger proportion of costs — or more specifically, where it represents a larger proportion of profit/profit margin. Factors such as the scope to raise finance against property and the pros and cons of sale and leaseback may come into the evaluation of the merits of business growth.

Business Skills for General Practice Surveyors only touches upon the running of surveying businesses, but the illustrations show the need for surveyors to be conscious of factors such as supply, demand, price sensitivity, monopolistic opportunity, competitive threat and the economics of client businesses. (See chapter 7 for more detail on company accounts and financial issues affecting surveyors' work.)

Market intervention and types of economy

A "free market" or "*laissez-faire*" concept is where government does not intervene in the market, such as through price controls or legislation. The other extreme is a "centrally command" or "planned" economy, which is run by government.

In a "mixed" economy, such as the UK, government intervention includes competition policy and regeneration initiatives, and a system of taxation and public spending that distributes income and wealth fairly.

Economic variables and economic policy

This section examines the main economic variables, and how the UK economy operates. It considers the basics behind economic growth, interest rates, inflation, taxation and public spending, and international trade and exchange rates. More detailed aspects are then considered.

Economic growth

Economic activity is measured by gross domestic product (GDP), which is the total annual value of the output of UK goods and services.

As with all nations, the UK government aims to secure "economic growth" and generally improve the living standards of its population.

Economic growth is measured as the increase in "real" GDP (ie net of inflation) and is assessed with reference to the longer-term average of economic growth, known as the "trend rate". In the UK, the Treasury estimates this at 2.75% pa but, for the purposes of fiscal projections, only calculates it at 2.5%.

Economic growth needs to be sustainable, with a country enjoying stable economic conditions. The opposite is boom and bust, which is disadvantageous in terms of uncertainty of businesses as to the economic outlook and the impact this has on their investment decisions. Examples include an overseas manufacturer considering against relocating to the UK, and a local manufacturer deciding not to increase output capacity through additional premises, plant and machinery, labour and so on. Similarly, uncertainty causes consumers to hold off spending. The reduced spending by businesses and consumers decreases economic activity and restricts economic growth. If the effects are severe, then a country goes into recession.

Interest rates and inflation

Interest rates are the principal tool used to control the UK economy. Since 1997, they have been set by the Bank of England's Monetary Policy Committee (MPC). This reflected the view of the Labour Chancellor, Gordon Brown, that interest rate decisions that were free from political interference would be more conducive to stable economic growth. The setting of interest rates is known as "monetary policy".

The MPC sets interest rates by considering the rate of inflation in the economy. The higher the rate of inflation, the higher interest rates will need to be. Although inflation has been relatively low over the past 10 years (generally in the region of 2–3% pa), in the 1970s and 1980s inflation rates of 10% pa, and higher, had a damaging effect on the UK economy. High inflation, like boom and bust, increases uncertainty. It also makes UK exports more expensive in overseas markets and therefore less competitive, thus depressing output, economic activity and economic growth (see "Exchange rates and international trade" below).

To bring about sustainable economic growth, the rate of inflation needs to be stable, and the management of interest rates helps achieve this. The government, and likewise the Bank of England's MPC, work to an inflation target. This is now 2%, based on the Consumer Prices Index (CPI), which represents the change in prices of a basket of goods and services. This recently replaced the Retail Price Index (RPI) as the target measure (see p82 for further detail). The 2% target rate of inflation is considered conducive to stable economic growth, with a small amount of inflation generally being regarded as a good thing.

As it is activity in the consumer sector that has a particularly sensitive effect on the rate of inflation, increases in interest rates help reduce consumer spending/consumer demand. Here, rises in interest rates increase the level of interest repayments on domestic mortgages, and therefore reduce the remaining income available for spending on goods and services. Other forms of credit also become more expensive, and therefore less attractive, and again help curb spending. The ability for consumers to obtain credit, and therefore the extent to which bank and other lending is relaxed, is also of influence (in the same way that the ability to borrow money influences the purchasing power for investment and development property, and therefore its price). As already mentioned in respect of supply and demand, reduced demand helps lower prices. Higher interest rates can also increase the preference to save instead of spend, although this is a relatively small influence.

However, while higher interest rates maintain the rate of inflation at the right level for the economy through the management of consumer spending, consumer spending still needs to be at a level that is conducive to economic growth. Management of the economy is therefore, to some extent, a balancing act. If, towards the end of a period of strong economic activity and economic growth, interest rates remain too high for too long, this may indeed subdue inflation, but can make a downturn more severe (ie a "hard landing" for the economy, rather than a "soft landing"). Likewise, if an economy moves strongly out of recession, delays in increasing interest rates to the necessary levels can fuel inflationary pressures.

Exchange rates and international trade

Exports are an important element of UK output, and the level of output is influenced by the "exchange rate" — also referred to as the "strength of the pound". The pound strengthens against the euro, for example, when the exchange rate moves from £1 = 1.4 euros to £1 = 1.5 euros — in other words, more euros can be purchased for a pound. Likewise, the pound weakens against the US dollar when £1 = $2 moves to £1 = $1.80 (because less dollars can be bought for a pound).

If the pound strengthens, and the exchange rate is high against foreign currencies, UK exports are more expensive in overseas markets. They are therefore less competitive, and UK output/sales are likely to suffer. At the same time, the high value of the pound makes imports cheaper — such as 1.5 euros worth of goods now costing 1.4 euros — or 1.5 euros worth of goods now being available for 93p instead of £1. Cheaper imports are able to better compete against the UK-produced goods and services, thus reducing UK companies' output/sales. A higher exchange rate therefore has the twofold effect on the UK of reducing outputs for both overseas and domestic markets.

A strong pound, because of the above processes, does, however, ease inflationary pressures in the UK. This is because reduced UK output decreases consumer spending/consumer demand — including via unemployment, and through reduced wages/spending power for those in work. It is also because imported goods and services are cheaper (including component costs of the manufacturing process as well as retail goods/services). Low UK output and high unemployment levels, however, have a detrimental effect on economic growth.

In contrast, a weak pound (low exchange rate) helps boost UK outputs, employment, consumer spending, economic growth and such like. However, together with the increased cost of goods and services that are imported, this can be inflationary. The above points show that a strong pound can be beneficial for the economy by lowering the cost of imported goods and keeping overall inflation in the economy low. As a result, the government/Bank of England could, in theory, allow growth to be above trend without creating inflationary pressures (such as by keeping interest rates low).

As a further illustration of the effect of interest rates in controlling the rate of inflation, increases in interest rates increase the strength of the pound/exchange rate, leading to the aforementioned effects. The pound strengthens because higher interest rates increase the returns available from depositing money in the UK, and consequently there is greater demand for the pound. The strength of the pound is also influenced by the success and stability of the UK economy, as this will be conducive to overseas investment in the UK. Exports, of course, also increase the demand for sterling, as does the flow of all money into UK financial markets (and likewise, imports and other financial outflows from the UK reduce the demand for sterling).

Another illustration of the issues involved in economic policy decisions is that the UK manufacturing sector may be weak, and therefore demand lower exchange rates (achieved through lower interest rates), while the UK service sectors/retailing sectors are strong, causing concern over inflationary pressures. Although manufacturing represents the majority of export trade, service/retailing sectors represent a substantial (and generally increasing) proportion of UK output/GDP. The exchange rate is also influenced by the performance of overseas economies, bearing in mind that the relationships outlined in this chapter for the UK will operate similarly in other countries.

Also, the "balance of payments" refers to the flow of money into and out of the UK. The balance is in "surplus" when the inflow exceeds outflow and in "deficit" when outflow exceeds inflow. The "current account" refers to flows of goods (known as "visible trade" or the "trade balance", with a deficit being termed the "trade gap"), services (known as "invisible trade") and transfer payments, which include foreign aid and military spending. The capital account refers to financial flows, such as investment in property, equities, gilts and cash/deposits.

Fiscal policy

Government policy in respect of taxation and public spending is known as "fiscal policy". If the government is collecting more income through taxation than is being incurred on expenditure, there is a "budget surplus". A "budget deficit" is where expenditure exceeds income receipts.

The Public Sector Borrowing Requirement (PSBR) is the level of debt needed by government within a fiscal year. This will vary depending on economic conditions, because if, for example, the economy is performing strongly, there should be a relatively high level of tax receipts and relatively low levels of social security spending in respect of unemployment. This is why, for example, when the government announces its spending plans, reference is also made to economic growth and the amount of taxation that a certain level of UK output will generate in order to meet spending commitments. The PSBR is also influenced by the government's attitude towards public spending — such as spending on health and education in the run up to an election without simultaneously raising taxes.

The government's total level of debt is the "national debt". This is financed by government borrowing (such as through gilt-edged securities or "gilts" market). As with all costs of finance/borrowing, interest payments/debt servicing adds to spending/PSBR.

Both public spending and taxation influence economic activity and economic growth. A reduction in taxation, for example, or an increase in public spending, will give consumers greater spending power. Therefore, if the economy was performing strongly, and tax cuts could easily be afforded, the government may still not make tax cuts due to their potentially inflationary effects. Instead, the PSBR can decrease and there may even be a budget surplus — ie a Public Sector Debt Repayment (PSDR) and a reduction in national debt. In a recession, tax cuts could help stimulate the economy, but would increase government debt.

Although public spending and taxation can be used to manage the economy, inflation etc, in such a way, interest rates are the primary tool. Interest rate/monetary policy is also a more flexible tool than waiting for budgets in order to adjust taxation. Also, much of public spending is on mandatory items, and it is therefore difficult for cuts to serve short-term economic needs.

Macroeconomics and microeconomics

Macroeconomics refers to the operation of the whole economy, such as through the setting of an inflation target and interest rates, and taxation and public spending. Microeconomics involves more detailed elements, such as support programmes for particular industries or competition policy.

Economic complexities and market dynamics

The simple relationships outlined above are, in practice, subject to many forces that make it difficult to manage the economy accurately and also to predict future trends. This is due to the various interrelationships between the variables, and also factors such as the accuracy of economic statistics. While historical performance of the economy helps indicate broad trends, circumstances and variables are forever changing, and previous occurrences are unlikely to be repeated with great consistency.

There are also lead and lag effects. A good example is the MPC's interest rate decisions having an effect on the expected inflationary levels in 18–24 months' time, as well as to the most up-to-date economic data available. This is because it can take time for inflationary pressures (and also deflationary pressures) to be reflected in price levels.

Speculative activity in the financial markets, including global financial flows, is also a factor, and could, for example, mean that exchange rates are at artificially high, or low, levels.

The psychology of consumers and markets plays a role too, with sentiment, confidence and comments in the media all influencing decisions to spend, save, invest, move house and such like.

The following sections illustrate, in more detail, the effects of the housing market on the economy, costs and other influences on inflation, and measures of inflation.

Housing and the economy

The performance of the housing market has a relatively sensitive effect on the economy. The broad level of house prices is influenced considerably by the level of interest rates and the cost of mortgage payments. Lower interest rates naturally make property more affordable and enable buyers to pay more by way of higher house prices. Higher income, through wage rates, lower taxation etc similarly adds to homebuyers' purchasing power.

The many other factors affecting house prices include:

- the lending policies of banks and building societies
- the strength of the residential investment market
- the popularity of second homes
- demographics and the increased number of single people in need of a home
- the number of new homes built per year (and existing homes demolished, ie slum and tower block clearance)
- any constraints on the potential to increase the supply of homes, including planning policies and the commercial viability of residential development.

The housing market is a good example of where the balance between supply and demand determines price, and an excess of demand over supply can increase prices rapidly. However, unlike many other goods and services that are purchased outright, price is a function of the affordability of mortgage interest payments and other housing costs. Therefore, price levels reflect a combination of primary supply and demand, and debt servicing and other costs.

The housing market is also a good illustration of how confidence and psychology can influence markets. Increases in house prices create a feel-good factor among existing homeowners, whose greater perception of wealth and financial security induces higher consumer spending. Increased prices also provide scope for homeowners to realise funds/equity through remortgaging, again leading to higher consumer spending, such as on larger consumer goods, cars and holidays. A high level of transactional activity in the housing market also leads to increased consumer spending on home furnishings and home improvements. As illustrated earlier in this section, such factors add to inflationary pressures in the economy. Increased equity, as a result of higher house prices, also adds to the spending power for a new, larger, more expensive home.

The market can be fuelled by overconfidence. Homeowners can stretch themselves to the limit financially and leave no margin in the event of unemployment or illness. The need to "keep up with the Joneses", and certain types of people being troubled by the impressions of others, adds to consumer spending. Finances can also be squeezed as interest rates rise and mortgage repayments increase. This will be the case particularly with variable rate mortgages, although fixed rates, discounted rates or similar, may provide a cushion for some time.

Confidence can, however, subside when interest rates begin to rise, and there are more frequent reports in the media of the end of the house price boom, and the risk of a property crash. Whereas one of the factors previously driving the demand for housing was its prospect of increasing in value, the fears of price falls can help halt demand, and the level of transactional activity, thus depressing prices.

Indeed, governments, the Bank of England and anyone else looking to influence markets will use speeches, articles and research to help sway markets in their desired direction — such as mooting the prospect of a certain level of interest rates by the end of the year, or reporting a particular economist's view that house prices will fall. This is similar in property circles to agents "talking up the market", and the sometimes perceived unreliability of agents' views and market research releases.

Cost side and other contributors to inflation

As already mentioned, inflation is influenced by economic activity and particularly the level of consumer spending in the economy. As well as such demand pressures, inflation is also influenced by cost pressures. Businesses sometimes report, for example, that higher costs "will have to be passed on to the consumer", ie through higher prices. Examples of costs include wage rates and the price of raw materials and commodities (including oil prices — see p87 for further detail).

As an example of interrelated effects, higher wage levels increase consumer spending power, as well as being an additional cost — ie creating both "demand-pull" and "cost-push" inflationary pressures.

The scope for demand pressures and cost pressures to increase the rate of inflation depend on structural factors within the economy and on economic capacity. Examples of structural factors include efficiency gains achievable from new technologies and improved working methods. Sometimes this will offset the effects of wage inflation. The flexibility in labour markets means that businesses can hire and fire with relative ease, and not be burdened with undue labour costs. The insecurity of labour helps prevent costs rising (as employees more readily accept lower wages and are reluctant to push for higher wages). In contrast, the power of trade unions can lead to wage increases and other restrictive terms of employment.

Where there is competitiveness within industries and among firms, there is less scope to increase profit by raising prices, and attention can focus more on cost control. High street retailing, for example, is highly competitive, with price transparency, and increased consumer demand does not tend to lead to unduly inflationary price increases (although there will usually be some effect). International competitiveness also helps ensure that domestic firms are put under pressure by overseas competitors to minimise costs, and generally enjoy less scope to increase prices.

Economic capacity is measured by the "output gap", which is the available capacity in the economy that can be used to increase production/output. If resources become scarce in supply, increased demand results in increased prices. Capacity can be provided by high unemployment but, more specifically, the availability of skilled labour. As has been demonstrated in the parts of the construction industry over recent years, an excess of demand against supply of certain skilled tradesmen has led to increases in its cost. Financial resources, property holdings and other assets add to capacity, and inventory levels provide a shorter-term influence. Previous levels of investment in the economy will be important, such as in respect of the training of skilled labour, and regarding the acquisition of capital assets.

Measures of inflation

Inflation is measured in a number of ways. This helps the different contributors to inflation to be examined, bearing in mind that some may have the scope to further accelerate the rate of inflation, whereas others may be temporary and/or relatively unimportant factors.

As mentioned above, the UK government adopts an inflation target, which, since December 2003, is 2% per year, and relates to the Consumer Prices Index (CPI). This derives from the European basis of measuring inflation of the Harmonised Index of Consumer Prices (HICP). CPI and HICP are calculated in the same way, but are of UK and European application respectively. HICP enables comparisons in the rate of inflation to be made across European states. CPI/HICP measures the monthly change in the price of goods and services purchased by households.

The government's previous inflation target was based on RPI-X, which is the Retail Price Index excluding mortgage interest repayments. The Retail Price Index, as with CPI/HICP, measures the price changes of a basket of goods and services, but differs in the households and goods and services represented, and how the index is calculated. The target for RPI-X was 2.5%, and the 0.5% difference with CPI is due to the different component parts of the indices, particularly the representation of a number of housing costs in RPI-X, not included in CPI.

Price levels in the commercial sectors are also measured. The Producer Price Index (PPI) covers the monthly change in the price of goods bought and sold by UK manufacturers, and the Corporate Services Price Index (CSPI) covers the quarterly changes in price of services provided by UK businesses to other UK businesses. There are various other indices, such as to measure import and export prices, and new indices are sometimes developed, such as to reflect new technologies or industries.

It is worth noting that the Bank of England only has operational independence. The government continues to set the inflation target and the inflation measure (and the Bank argued against the government's change from RPIX to CPI). In theory, the government could ask the Bank to target inflation at a higher or lower rate than 2%.

Economic and Monetary Union

Economic and Monetary Union (EMU) is the use of the same, single currency — the euro — by member states of the European Union (EU).

The UK is a member state of the EU (since joining in 1973), but has not joined the single currency. Denmark and Sweden are the other members of the EU but not EMU. EMU members are Austria, Belgium, Finland, France, Germany, Greece, Ireland, Italy, Luxembourg, the Netherlands, Portugal and Spain. The 10 countries joining the EU in 2004, none of which are yet part of EMU, were Cyprus, the Czech Republic, Estonia, Hungary, Latvia, Lithuania, Malta, Poland, Slovakia and Slovenia.

Whether the UK will join EMU is dependent on both economic and political issues. In order for economic conditions to be suitable for the UK to join EMU, there needs to be compatibility with the European economy — known as "convergence criteria" or "Maastrict criteria", as they were set out in the Maastrict Treaty. When EMU commenced in January 1999, members had to meet criteria relating to interest rates, exchange rate stability, inflation, government borrowing and national debt — but not unemployment and economic growth. New members similarly have to meet the criteria. Similar ongoing requirements for members of EMU are provided in a "Stability and Growth Pact".

A single interest rate applies to the members of EMU, set by the European Central Bank, which is responsible for European monetary policy. A single monetary policy aims to secure low inflation and price stability because, as with the comments on the UK economy in the previous section, such conditions are conducive to stable economic growth, increased international trade and lower unemployment.

An illustration of the incompatibility between the UK economy and the European economy was shown in January 1999, when UK interests rates were 6.25% and European interest rates were 3%. By December 2004, UK rates had fallen to 4.75% and European rates were 2%.

If UK rates fell to European levels, severe inflationary pressures would emerge, led particularly through increased house prices. The UK economy tends to be more sensitive to interest rate changes because it operates on higher levels of debt, and a greater proportion of variable rate (as opposed to fixed rate) debt than other European countries. The UK economy is also more exposed to exchange rate variations, as it is more open to international trade than European countries.

Membership of EMU would mean the UK government would lose its ability to control the economy through domestic interest rates. If, for example, the UK economy was performing strongly while the rest of Europe was experiencing a relative downturn, interest rates could not be increased in order to control inflationary pressures. Likewise, interest rates could not fall in order to lessen the severity of UK economic difficulties. The UK would also be reliant on the euro as an exchange rate, and its strength against currencies elsewhere in the world — as opposed to the exchange rate being against the euro as well as other currencies. Although the UK economy could be managed by fiscal policy (taxation and public spending), this is not as flexible as the effect of interest rates, and there are also restrictions provided by the Stability and Growth Pact.

Benefits of joining EMU include a growth in international trade, increased international competitiveness of Europe against the rest of the world, and increased price transparency (and therefore reduced rates of inflation). If the UK did join EMU, it would be important that entry was at an appropriate exchange rate (ie the strength of the pound against the euro), as this would be irrevocably fixed. In theory, the UK would have to participate in a new exchange rate mechanism, ERM II, for two years prior to entry — but the Treasury is insisting that this will not be the case.

The government has set five economic tests relating to the UK joining EMU:

1. Are business cycles and economic structures compatible so that we and others could live comfortably with euro interest rates on a permanent basis?

2. If problems emerge, is there sufficient flexibility to deal with them?
3. Would joining EMU create better economic conditions for firms making long-term decisions to invest in Britain?
4. What impact would entry into EMU have on the competitive position of the UK's financial services industry, particularly the City's wholesale markets?
5. In summary, will joining EMU promote higher growth, stability and a lasting increase in jobs?

For countries to live on a sustainable basis within a monetary union, there needs to be a high degree of market flexibility. Economic theory states that countries need to have flexibility in real wages (wages after taking into account inflation), so that competitiveness can be improved if the option of devaluation does not exist. Also, labour needs to be mobile, and able to move to locations in the monetary union where the labour market is strong. In Europe, there is currently a lack of flexibility in the labour market and mobility, which has contributed to high unemployment in many countries.

Economic commentaries: *RICS Economic Brief*

RICS *Economic Brief* provides an overview of economic issues and comments on the commercial and residential property markets, as well as the construction sector. As the *Economic Brief* is published monthly, it is ideal for surveyors who need to remain abreast of broad economic trends. European economic issues are included in a similar RICS *European Alert*. Details are available via the RICS website *www.rics.org*.

In order to put some of economic variables and economic issues into practice, an extract is provided from RICS *Economic Brief* of May 2004. This was prepared by RICS Chief Economist, Milan Khatri.

The Bank of England (BoE) raised interest rates for the third time in seven months in May by 0.25% to 4.25%. A strong labour market, surging high street sales and improving business activity both at home and abroad, were used as a rationale for increasing the cost of borrowing. Concerns over inflation are to some extent justified given the recent rise in the oil price to record levels, though price pressures continue to be mitigated by a relatively strong sterling exchange rate. With underlying wage growth also still subdued, it seems that the BoE's pre-emptive interest rate hikes reflect mainly a desire to forestall volatility in house prices and the economy. In the near-term the impact on the economy will be muted as the cost of servicing debt remains low on an historical basis, whilst further interest rate rises will only be gradual.

The tone of economic data over the last month has been generally upbeat. Most encouraging were labour market figures showing employment in Q1 rising an annual 1.1% compared to only 0.5% in Q4 2003. Notably, full-time employment which had weakened in the latter part of last year, showed further signs of revival, up 0.6% over the year compared to a flat position at the end of 2003. The convincing turnaround of the labour market has led the BoE to becoming more hawkish on inflation prospects, wary of the impact of a tighter labour market on wages.

Wage growth accelerated to 5.2% in the first quarter, the highest pace since mid 2001. However, pay has been boosted by large bonus payments in the financial sector. On an underlying basis, wage growth is still muted though has picked up to 3.9% in Q1 compared to a recent low of 3.4% in Q2 2003. Wage rises are still below the 4.25% rate which is viewed as consistent with inflation at the government's target rate, but is likely to pick-up as employment conditions tighten, not least because of strength in government recruitment to boost the provision of public services.

One reassuring aspect for inflation-watchers is that the ongoing expansion in employment over the past year has to a large extent been driven by part-time workers. Growth in this category accounted for two-thirds of the increase in jobs with much of the rise in the form of self-employment. Wage inflation is traditionally not driven by this type of employment, and therefore poses less of a risk to the wage outlook over the near-term. Nevertheless, recent business surveys show improving economic conditions.

The Chartered Institute of Purchasing and Supply (CIPS) monthly survey of service sector firms continued to report sharp growth in business activity in April, which is close to the four year high seen in January. This bodes well for service sector employment growth in the months ahead and full-time employment positions. However, the service industry recovery has been driven by the financial sector, which might be affected by recent stock market weakness, particularly if it is sustained.

For the manufacturing sector, the CIPS monthly survey suggested that activity accelerated again in April, having eased in the previous month, growing above its long run average. Survey figures generally have shown strong growth in manufacturing activity over the past 3–6 months, and will have helped to sway the BoE's decision to raise interest rates, even though official data point to surprisingly modest growth in the overall economy for the first three months of the year. GDP rose 0.6%, which is only around the economy's trend rate of growth compared to 0.9% in Q4 2003. Moreover, manufacturing figures released after the BoE's interest rate meeting suggest that the sector is worryingly weak.

Manufacturing output fell 0.3% in March, with the trend over the last three months pointing to a renewed deterioration. The data is at odds with a host of business surveys, with the latter finding more weight in the BoE's thinking due to the tendency of initial growth estimates to be revised upwards. This also extends to the sector breakdown of the production sector, where the Bank has ignored softness in capital goods output and imports which are potential pointers to weaker business investment. Instead, improving corporate profits and margins give the BoE confidence that a sustained recovery in business spending and exports will endure, particularly as world economic activity accelerates.

The turnaround in manufacturing (as indicated in surveys) together with ongoing strength in retail sales, suggest that a broad-based economic recovery is under way. Retail sales rose an annual 6.6% in the first quarter, its strongest for two years after sales picked up again in March. The annual rate of growth has been partially boosted by subdued spending in the period running up to the Iraq war, though there has been substantial strengthen in spending this year. Nevertheless, given the robust state strength of the housing market and the growth of equity withdrawal, one would have expected consumer spending to be even stronger.

The traditionally tight relationship between house prices and consumption has weakened. This suggests that factors which are driving consumption, such as expectations for future income and job prospects are not present in the housing market to the same extent. The divergence, which is identified by the BoE in its May "Inflation Report" is probably explained by the rising influence of investment demand for residential property, as households have switched savings from the equity market into alternative assets, partly in response to concerns over future pension values. As such, a slowdown in the housing market driven primarily by slower investor demand need not necessarily provoke a strong reaction from consumer spending.

Aside from the housing market, the price outlook is quite benign on the domestic front, with inflation in March within touching distance of the 1% level that automatically triggers a letter of explanation from the Governor of the BoE to the Chancellor. Inflation though is likely to rise in the months ahead because of several external factors, not least the recent jump in the price of oil. Manufacturers' raw material costs in the year to April rose 3.4% due almost entirely to higher crude oil prices. The impact of higher oil prices is not only a concern for inflation but also economic activity.

If the high oil price level is sustained then economic growth will be curtailed as high oil prices act as a tax on consumers and businesses. Moreover, oil prices may rise further because of the turnaround in the global economy. Specifically the US recovery together with the growing presence of China as a major oil consumer is raising oil demand. China was the second largest oil consumer in 2003 after the US, the first time on record. The other key issue which is pushing prices higher is grave uncertainty over the situation in Iraq and the Middle East generally, such as recent attacks on oil installations which could hit supplies.

In its Inflation Report, the BoE pointed out the diminishing importance of oil for inflation as the weight of goods in the cost of living for households had declined over time. However, while the effect of the current spike may fade, higher oil prices look set to be maintained over the medium term. The 5.5 million barrels China consumed in 2003 may pale in comparison to the US's 20 million, but this represented a 77% increase in 4 years. Given its rapid rate of industrialisation, oil consumption in China is set to continue growing at a frenetic rate, meaning its influence on the oil price is unlikely to be one that goes away anytime soon.

Set against the rising oil price, is the subdued price picture in the rest of the economy. Tough competitive pressures have kept final prices to households and consumers under control, with the core inflation rate for factory gate prices easing back in recent months. So although there may be some genuine worries over oil prices, the data does not suggest there is an underlying pick up in inflationary pressures, despite notable economic strength. A rising exchange rate over the last 12 months has mitigated cost pressures, with the price of imported goods falling 2% below year ago levels in the first quarter following a brief rise at the end of 2003.

But rising oil prices will lift inflation in the next year or so and also lead to weaker economic growth in 2005. The BoE will tolerate a rise in inflation which arises from one-off factors, though it is the risk that oil price rises feeds into general wage and price pressures that is a main concern, particularly at a time of a tight labour market. We think that interest rates will continue to rise at their current gradual pace, reaching 4.5% in August and 4.75% in November — with a dampening of the housing market a key target because of the fears that any potential market bust may significantly damage the economy. However, we do not expect a step-up in the pace of rate rises, given the volatile situation in the Middle East and the recent fall out in the stock market. The economic impact of higher interest rates is unlikely to be felt until later in the year and 2005, particularly as debt servicing costs for businesses and consumers remain relatively low on an historical basis.

Property and the economy

Chapter 2 provided illustrations of the relationship between the economy and property market performance, with particular emphasis on market trends. A more detailed illustration is set out below, comprising an extract from the review of the commercial property market in RICS *Economic Brief* of May 2004, again prepared by RICS Chief Economist, Milan Khatri.

A strengthening economy signals that nominal rents are set to start rising this year, but a recovery in real terms is only likely by the end of 2005. The stability of the most recent rent cycle has surprised investors, following the acute volatility of the previous three decades. As such, investors are likely to continue putting money into the real estate sector, particularly given its strong performance compared to equities.

We expect real rents to only bottom out in mid 2005, though the decline from peak to trough will be around 3% in total over a 3-year period. This compares with a fall of 33% over a 6-year period in the early to mid 1990s according to Investment Property Data (IPD). CB Richard Ellis data suggest real rents fell 40% between 1989 and 1994, which is similar to the downturn of the early 1970s. Moreover, real rents are still 15% higher than the trough of 1994. This also contrasts with the early 1970s and 1990s, during which periods real rents ended lower than the initial starting point of the boom or had given back the majority of previous rises.

The stability of real rents is a reflection of good demand conditions, which has absorbed a strong rise in new building output. Private commercial and industrial construction output peaked at 1.8% of GDP in 1999, slightly below that of the previous cycle high of 2.0% in 1990. However, for the three-year period either side of both peaks, the average rate of building was 1.2% of GDP. The main difference between the two periods is that the most recent cycle has been fairly flat. The late 1980s/early 1990s cycle saw a classic sharp peak and then a collapse. The extreme rise in building in the late 1980s is likely to have exacerbated the subsequent downturn as newly completed property came onto the market during a period of weak demand.

As the global economic recovery is gaining traction, with a long awaited rise in US employment apparent in early 2004, it augurs well for growth in the world's largest economy and points to better demand prospects in the UK. As such, real rent rises are quite possible again by the end of 2005. The key factor underpinning rents will be stronger increases in employment within the private sector. The UK construction cycle alone is unlikely to prevent rents rising though will help to moderate the upturn. We expect nominal rents to rise 0.9% in 2004 and 1.3% in 2005, compared to a 1.2% fall last year. The risks to rents are on the upside as global growth is showing an acceleration according to survey evidence for both the service and manufacturing sectors.

The comparative stability of the real rental cycle in contrast to previous decade's volatility has surprised investors. Investor valuations of the market worsened (pushing yields higher) for almost two years in the aftermath of the stockmarket crash. However, pre-emptive monetary policy loosening, together with reflationary government policy has limited falls in rents, giving investors the confidence to re-enter the commercial property sector. Geo-political uncertainties and fears over the value of pensions have also enticed investors into the sector with yields in early 2004 down to the lowest level in 14 years.

Last year, the market was buoyed by foreign and private investor demand, with the latter using low interest rates to get a foothold in the market. Property Data statistics suggest that overseas and private investors accounted for 40% of all purchases in 2003 up from 31% in 2002. The share of private individual buyers has rocketed from 2% in 2000 to more than 10%, helping to drive property yields down. Borrowing costs are set to rise into 2005 due to a fully-fledged global recovery, removing some of the positive stimulus to demand. Investor interest for property may also wane if the equity market shows signs of a sustained upturn following strong outperformance in 2003.

However, the performance of the commercial sector is still substantially better over a 10 year period than for equities, giving investors a 174% return compared to 81% for the latter. Also, volatility of returns has also been much less, though this is partly because the UK has not experienced a recession over the period. As such, we expect investor interest in real estate to remain firm, nudging yields lower and capital values up as funds continue to diversify away from equities. Total returns for commercial property will rise to 11.5% in 2004 compared to 10.9% last year. Looking ahead to 2005, both higher interest rates and slightly slower economic growth will lead yields to stabilise, curtailing the returns on commercial property to less than 10%, but still remain positive.

RICS also publishes a series of "economic forecasts", which assess the prospects for the commercial property, residential property, construction and rural sectors. RICS "market surveys" similarly provide independent research across property sectors.

RICS Foundation commissions research in the property sector, which is often available free of charge, or at nominal cost. Research findings can help support investment and development strategies and decisions, as well as adding to the information on which surveyors may draw as part of their various property dealings. As an example, the summary from the paper *Benchmarking Urban Regeneration* states:

Without adequate information, investors are unwilling to allocate funds to a project or an investment opportunity. There are many areas of our towns and cities that need urban regeneration, which are being deprived of much-needed investment because investors are unsure of the returns that they could expect. This research has developed a property performance index for urban regeneration areas. The results show that it is possible to achieve financial returns that exceed recognised performance benchmarks. This finding is important in changing perceptions in investment decision-making and in the encouragement of private finance into urban regeneration.

Further details of the Foundation's work are available at *www.rics-foundation.org*.

Other market information

There is a wealth of market information available to surveyors. Daily newspapers include economic commentaries and weekend newspapers often provide more detailed analysis. *Estates Gazette* includes periodic comment from guest contributors, and particularly newsworthy items are covered in national television and radio news.

The research departments of the large property consultancies provide updates, much of which is available free of charge via their websites. Also, many of the high street banks produce similar market research publications.

The Bank of England website (*www.bankofengland.co.uk*) provides a range of information, including interest rate decisions, minutes of MPC meetings, inflation reports and other economic information. More detailed statistical information is available via the Office for National Statistics website (*www.statistics.gov.uk*), extending to many other areas of government. Finally, information from the HM Treasury is available at *www.hm-treasury.gov.uk* and the Office of the Deputy Prime Minister's website, *www.odpm.gov.uk*, may also be useful.

The Equity Market

The main element of the financial markets in which surveyors are likely to have an interest is the equity market. This chapter outlines its basic workings, and the relationship with the economy and property market. As well as being of relevance to surveying work in practice, the chapter should also be of interest to surveyors considering trading on the stock market. (Also, in view of the prospect of Real Estate Investment Trusts (REITs), known also as Property Investment Funds (PIFs), which would securitise properties/portfolios and enable their shares to be traded, surveyors' knowledge of the equity market will be important.)

Equities

Equities, known also as "shares", are held by "shareholders" in companies. Small private limited companies have shares and shareholders but, in the context of the equity market, shares are listed on the Stock Exchange, and are available to the public to purchase. These are really "ordinary" shares.

Shares are issued by companies in order to raise money, and investors/ shareholders will hope to gain income through "dividends" and an increase in the share price. This is similar to property investors requiring income through rent, and an increase in the capital value of their property.

However, a feature of equities is their greater risk compared with property — especially if holding equities in a single company or in a small number of companies. The worst-case scenario for an investor is that the company goes bust and the investment becomes worthless — ie the capital committed to purchase the shares is lost. Even when a company survives, shares prices can fall rapidly, and may take many years to recover — and some never reach their previous share price highs.

Shares are likely to play a part in the personal finances of surveyors. Pension funds, for example, in ensuring that longer-term liabilities can be met when pensions are drawn, invest heavily in equities. When there are reports in the media regarding problems in the level of pensions available when people retire, part of this reflects the level of the equity market and the value of shares. Surveyors may have invested savings in a "maxi ISA", which includes shares, or they may have a small portfolio of shares as part of their personal financial planning. As an interest, some surveyors may actively trade in shares, either alone or as an investment club set up with colleagues or family/friends.

FTSE 100 index

The FTSE 100 index, which is commonly referred to in the news and economic and property market commentaries, comprises the 100-largest companies whose shares are listed (or "quoted") on the Stock Exchange. Examples include Boots, British Land, BP/British Petroleum, British Telecom, Dixons, HSBC, Land Securities, Lloyds, Marks & Spencer, TSB and Vodaphone. Listings are shown in daily newspapers, with the most comprehensive being in the *Financial Times*. This also shows the sectors within which companies' shares are listed, such as banks, general retailers, oil and gas, real estate and telecommunication services.

"Largest" refers to the capital value of the company or "market capitalisation". This is the share price multiplied by the number of shares in issue. The FTSE 100 index is a weighted average of market capitalisations and, in June 2005, was in the region of 5,000.

Changes in membership of the FTSE 100 index reflect the growth of companies, the decline of companies and also mergers and acquisitions. In calculating the value of the index, dividends are ignored, as is the effect of inflation.

Although the term "blue chip" is commonly used to denote large companies with strong financial standing (or "covenant strength"), FTSE 100 companies may not necessarily be financially sound. Dotcom-type companies, for example, have become members of the FTSE 100 index due to their high share price based on future expectations, despite not having made any profits. Today, FTSE 100 companies can go bust. Other indices include the FTSE mid-250 index, which comprises the 250 next-largest companies. Examples include Matalan, Slough Estates, Stagecoach and WHSmith.

Share prices

The share prices of individual companies on the stock market range from around 1p per share up to £10 per share, although some are higher. Shares in the FTSE 100 companies, however, will generally be above 50p. By convention, share prices are quoted in pence, not pounds.

When share prices are quoted in the financial press, this represents the "mid-price" between the price at which the shares can be purchased (the "offer" price — ie the price offered to investors/purchasers to buy) and the price at which shares can be sold (the "bid" price — ie bid to investors/owners to sell).

For shares in the larger companies, the "bid-offer spread" will be relatively narrow, reflecting the frequent trading and liquidity in the market for the shares

— such as 248p to sell and 250p to buy. For smaller companies, the prices could, for example, be 20–22p, broadly meaning that the shares will have to increase by 10% before the investor breaks even (ignoring costs). In the case of "penny shares", which generally comprise small companies whose share prices are denoted at only a few pence, and which may provide high speculative returns (or just as easily lose all value), prices could be 3–4p — thus having to increase by 33% in value before breaking even (ignoring costs). Another effect on liquidity and price is the "normal market size" of individual transactions (ie the number of shares being traded). For small companies, this is low, and may involve a small premium having to be paid to the above prices if trading amounts to above the normal market size — such as the 20–22p share having to be purchased at 23p, or sold at 19p.

The financial press also usually quote the high and low share prices for each company. Depending on the time of year, this will be based on the prices in the current calendar year, or the prices during the past 52 weeks.

The shares of the larger companies trade by way of an "order-driven" system, which means instructions to buy and sell are entered into a central computer system at the required buy or sell price, and then matched automatically. Other shares are traded by a "quote-driven system", which involves "market makers" quoting the price of shares for buyers and sellers to then agree whether to trade.

The share price of an individual company is determined by many factors, but of primary importance are its earnings/profits, the amount of dividend paid to shareholders, and the company's prospects. This includes the scope for increased dividends, with the potential for share price increases also being important.

Economic factors

Economic factors are another important influence on share prices and, as mentioned in chapter 5, include the rate of inflation, interest rates and the exchange rate. Shares perform best when there is low inflation, low interest rates and rising company earnings.

Increases in interest rates generally create downward pressure on share prices. Cash holdings can then secure a higher return, by way of interest payments, compared with equities, and this increases the relative demand for cash deposits against investing in shares. Higher interest rates are also likely to mean lower company earnings, because economic activity and consumer spending subsides, and also because finance costs are higher — which may limit investment expenditure, thus restricting expansion and growth prospects. However, as an increase in interest rates is likely to be due to strong levels of economic activity, investors may not be too concerned with small interest rate rises because of the beneficial overall effects of economic activity on company earnings.

Sometimes, an increase in interest rates may even be welcomed by the financial markets — such as where a pre-emptive strike is taken in order to prevent interest rates being higher in the future, and the economy generally becoming destabilised. Certainty and stability are key factors for the equity markets, with investors typically discounting disproportionately highly to account for uncertainty and risk. Stable economic activity and low interest rates encourage investment, thus bringing economic growth and an increase in company earnings.

An interest rate rise could also increase share prices because it is actually less than the increase that had been feared, or simply because an announcement ends the speculation and uncertainty that had hitherto caused share prices to be depressed.

Different sectors and different companies will be affected by changes in interest rates in different ways. Companies that are heavily reliant on debt, for example, will be more exposed to interest rate changes, and luxury goods companies are more likely to suffer if consumers' disposable income is squeezed, compared with the relatively stable demand that retailers of essential goods may experience.

Because, as outlined in the previous chapter, interest rates are primarily influenced by inflationary pressures in the economy, the financial markets monitor statistics in a similar way to economists and the Bank of England's Monetary Policy Committee.

As also mentioned in chapter 5, the exchange rate, and the "strength of the pound", influences the pricing/competitiveness of imports and exports, and therefore the earnings of companies involved in international trade. This is particularly likely in the case of FTSE 100 companies.

The performance of other countries' economies similarly influence their stock markets and, likewise, the performance of overseas stock markets influences the UK stock market. The relationship is particularly sensitive between the Dow Jones Industrial Average in the US (at Wall Street), especially where there are large price changes. "The Dow" is the equivalent to the UK's FTSE 100 index, but includes only 30 companies, such as American Express, Coca Cola, IBM and Wal-Mart.

A "bull market" is where shares are performing well, and a "bear market" is where they are performing poorly. At June 2005, the market is relatively static.

Other asset classes

The performance of the UK equity market is also influenced by the returns available on other asset classes. This is because institutional investors hold a mixed portfolio of assets, including equities, gilts/bonds, property and cash, and look to diversify risk. In contrast, property investors are likely to be exposed solely to property, with the management of risk achieved through a variety of property types, locations and investment profiles (as well as through sensible levels of borrowing/gearing).

Gilts, which are government-backed bonds, provide a fixed rate of return in relation to the initial capital outlay. A gilt, when issued at £100, may provide a return of £5 per year, equating to a gilt yield of 5%. £5 will always be paid, but the price of the gilt will vary, and therefore determine the gilt yield at any point in time. If, for example, the price rises to £110, the gilt yields 4.5% (5 ÷ 110). The gilt will have a redemption date, at which point £100 is returned to the investor.

"Long-dated" gilts, which have over 15 years to maturity, have larger price movements, because they are less influenced by the value at redemption — ie a short-dated gilt, which has less than five years to maturity, priced at £120, will lose £20 value over the period to redemption. When property yields are being compared with gilt yields, long-dated gilts are used because this shows yields that are not influenced by the imminent redemption.

Gilts perform well (in terms of increasing price/capital value) when the rate of inflation is low, economic activity/growth is low and interest rates are low. Gilts can also be in demand when the equity market is struggling and/or experiencing particular uncertainty, and investors seek a "flight to quality". Gilts are also influenced by the government's finances, as an increased Public Sector Borrowing Requirement is likely to warrant an additional supply of gilts, thus depressing prices and increasing yields.

Earnings and other factors

A company's earnings are known as earnings per share (EPS) — ie total earnings divided by the number of shares in issue. This is necessary because the number of shares in issue can vary (see later). More specifically, "diluted" EPS is considered, which accounts for share options which could be exercised, and "preference" shares which are convertible to ordinary shares (both of which increase the number of shares in issue and therefore dilute EPS). Preference shares provide a fixed rate of return/dividend in preference to the returns/dividends distributed to holders of ordinary shares.

The "price-earnings ratio" (P/E ratio) is the market capitalisation of the company divided by its earnings (or the share price divided by EPS). If the share price is 200p and EPS is 10p, the P/E ratio is 20. The P/E ratio is the principal means by which equities are valued (see also "net asset value" later). It is similar to how surveyors use years' purchase in investment valuation to multiply the rent. The property market, however, adopts "yield" for comparison and methodology. As an illustration in line with the above figures, a capital value of £2m and a market rent of £100,000 equates to a 5% yield and a years' purchase/multiplier of 20. In the case of the valuation of licensed and leisure interests, only a years' purchase/multiplier will be applied to a company's profits.

The higher the P/E ratio, the greater the prospects are generally considered to be for the company in the future. As is outlined later, many other factors influence the company's share price. A high P/E ratio is not necessarily good or bad: shares at any P/E ratio/share price could be undervalued or overvalued. P/E ratios for the larger quoted companies may generally be around 10–15, although may be 20 or more where there are particular growth prospects. However, as a further illustration of the need for close examination of a company's financial situation (including trends over recent years) a high P/E ratio may be due to particularly low profits in the current year, reflecting one-off factors. Conversely, profits may be exceptionally high in an individual year, and a P/E ratio therefore low.

Any surveyors interested in share trading will be searching for shares that are undervalued by the market and have scope for price increases — especially if looking for medium to longer-term returns, bearing in mind that the market can be slow to reflect the full potential of a company in its share price. Despite the theory that the stock market provides an efficient mechanism for accurate pricing/valuation, in practice shares can remain overvalued or undervalued for relatively long periods.

Surveyors who trade in shares may still buy shares that are overvalued because further share price increases are expected. The "dotcom bubble" in the late-1990s/early 2000 was a good example of how share prices can rise dramatically, irrespective of fundamentals — reflecting factors such as investor naivety and greed (with other terms used by financial commentators including "herd instinct" and "mania"). Psychology always has an important influence on the financial markets.

Some companies may have no profits/earnings on which a P/E ratio can be established, thus making it impossible to price shares with any accuracy. Therefore, when a company makes losses, or very low profits, its share price cannot be underpinned by a P/E ratio, and the share price may fall to a particularly low level.

Share prices can also fall sharply when a company's strong growth prospects falter — such as a company with a P/E ratio as high as 40 seeing profits/earnings increase by only 5%, causing the company/share price to be rerated. As well as an instant share price fall on the release of the earnings figures, progressive share price falls may leave the share price at, say, 50% or less of previous highs. It is not uncommon for some companies, although more so for those outside the FTSE 100, to see their share price at 10% of its previous high — ie the share now being worth 10p and not 100p.

The P/E ratio also helps companies to be compared with similar companies within their sector, although this is a better guide for companies/sectors with stable earnings. A P/E ratio is also adopted for the market as a whole, or for the FTSE 100 index. This helps assess the outlook for the market and whether its current level appears high or low.

The direction of the stock market and the potential extent of rises and falls, however, are notoriously difficult to predict. As with the illustrations in respect of the economy in chapter 5, and as explained below, there are a multitude of interrelated influences on companies' share prices, and on the performance of the market as a whole.

Dividends per share and dividend yields

"Dividends per share" is simply calculated by the total number of shares divided by the total dividends paid out. The "dividend yield" is the dividend payment divided by the current share price — such as 2% where the share price is 200p, and a 4p dividend is paid. Dividends include a final dividend paid at the end of the financial year, and an interim dividend paid midway. A "special" dividend is also sometimes paid.

Dividend yields on shares are relatively low, and are generally less than the interest rate return available on cash deposits, the yield available on gilts and the yields available from property.

A high or low yield is not a particularly reliable indication as to whether a company is doing well or not. A low yield could be a reflection of the growth potential of the company, and the prospect of dividends being higher in the future — and also the scope for increases in the share price. It is the scope for increase in the capital value of shares, particularly in the short term, which is the principal driver of share prices (notwithstanding the relationship between income/ dividends and share price). A low yield would also result from a company's decision to pay only a small proportion of its earnings out in dividends, and to invest in expansion instead. A high yield could be due to falls in the company's share price, reflecting its poor future prospects.

If, however, dividends are rising year after year, this would suggest that the company is doing well. As a further example of the need for detailed consideration of a company's financial information, if a company runs into difficulties, it may maintain (or even increase) the dividend in order to suggest that a downturn in earnings is only temporary. In the case of a struggling company, a high dividend yield can help support the share price.

"Dividend cover" is earnings divided by the dividend, such as 2.5 if, in the above example, dividends were 4p per share, and earnings were 10p per share. This helps indicate factors such as a high payout in difficult times, or the retention of earnings for expansion. Reference is also made to the earnings yield, which is calculated as EPS divided by the share price (or the total earnings divided by the market capitalisation), and is the inverse of the P/E ratio.

Other share price influences

The performance of the equity market, and in particular the price of individual shares, can be influenced by many unpredictable and fickle factors. This reflects the speculative activities of investors and the psychological nature of the markets.

If, for example, a company reports good results, the market may already have reflected these in the share price. If the news does not lead to a buying of shares and a consequent price rise, investors may sell the shares (or "take profits") and the share price will fall. In another situation, the market may not have been alert to the potential for improved profits, and good results lead to an initial increase in the share price, followed by more progressive increases as more investors slowly become aware of the potential. Whether good results are priced into the share price or not often depends on the size of the company, and the level of scrutiny it receives from the market (including the press).

Sometimes bad news will not be reflected sufficiently in a share price fall and, instead, the share loses value more progressively over time. Some companies may progressively leak bad news to the market in order to avoid a single shock that has a disproportionately severe effect on the share price. Bad news can often get worse and a share price can spiral downwards. However, in other situations, adverse fortunes can be temporary, and share price falls can be overdone, giving investors good opportunities from such "turnaround" situations. As indicated above, picking the likely highs and lows of the market or share prices is notoriously difficult — however, if a share price does not fall any lower on bad news, or any higher on good news, it may have reached its price limit for the time being.

As an example of the effect of speculative activity in the market, "short-selling" is where investors sell shares in companies where the share price is expected to fall, but without owning the shares. The shares are acquired later, at a lower price, thus providing a profit. This potential extra market of sellers (ie not just existing shareholders) can account for large falls in a share price — although the share price can still rebound relatively strongly when short sellers have to buy the shares.

Share price movements can be caused by comments in the press, and there have been reported cases whereby journalists have written favourably about companies whose shares they hold (and may have recently acquired). Brokers' or analysts' recommendations to buy or sell shares can influence share prices, and there are also cases of brokers making positive comment to the market in support of a client company, for it to be revealed that the actual opinion and advice to private client investors is the opposite.

Takeover and merger activity can cause share prices to increase considerably — and it is often the initial rumour in the press, or a formal announcement that has the most substantial effect on the share price, rather than the actual conclusion of the deal. Some suggested deals, however, see a share price fall because the market does not consider there to be any advantage in a deal. Indeed, there are examples of takeovers and mergers that have proved unsuccessful — some seemingly driven largely by the enhanced exhilaration felt by senior personnel and the chance to control a larger empire. Focus can be lost on the core running of the business, and it may be difficult to integrate parts of the new business.

If the share price of a company in a particular sector moves, the share prices of other companies in the sector can change too — such as because earnings growth, or decline, is expected to be similar. When a company changes sector, the share price can change, especially when sectors have recognised P/E ratios. If a company moves into or out of an index, its share price can change — such as a declining FTSE 100 company experiencing relatively large share price falls leading to its exit, owing to the many investors requiring FTSE 100 investments having to sell.

Some companies attempt to smooth out earnings variations from year to year in order to show steady growth and certainty — as this will be conducive to greater share price stability. This can be achieved through trading activities and certain legitimate accounting approaches. In other situations, however, and which is more likely with smaller firms, a company's profits may gyrate from year to year, creating opportunities for directors and any associates to trade profitably in the shares. Insider trading, scandals and other frauds are sometimes reported in the press, and serve as a further illustration of the care that has to be taken in assessing the true financial position and future potential of companies. "Pump-and-dump" denotes the promotion of shares by a company/investors, and "ramping" of the share price with a view to offloading the shares to unwary investors at overinflated prices.

As investors are warned, "the value of shares and other investments can go up as well as down", "past performance is not a guide to future performance" and "investors may not get back all or any of the amount originally invested". But, despite the risk and volatility of share prices and the stock market, it is often commented that, over the long term, shares provide superior investment performance to other forms of investment. However, in 1999, the FTSE 100 index reached highs of around 6,900, but in early to mid-2004 was only around 4,500 — a reduction of approximately 35%, with an increase of over 50% in the FTSE 100 index now being needed in order reach its previous high. At the time of the Iraq war in early 2003, the FTSE had fallen to around 3,300.

Other issues

As indicated above, the number of shares in issue, and therefore EPS, could vary. This could be due to a "rights issue", which is where a company issues further shares in relation to existing shareholdings, with shareholders being able to retain or sell their allocation. A company may also "buy back" some of its shares, such as where it has surplus cash that is not required for investment in the business. A rights issue therefore increases the number of shares, and reduces EPS. A buyback decreases the number of shares, and increases EPS. A buyback is preferable to shareholders when the share price is low — and companies may take advantage of a low share price to do this.

A "scrip issue" or "capitalisation issue" also changes the number of shares in issue, and is when, for example, a company has 50 million shares at 900p each, and issues a further 50 million shares to the shareholders (ie one share becomes two shares). This results in 100 million shares at 450p each, and no difference in the total value of the shareholding to the shareholder (with the payment of dividends, and dividends per share, similarly being adjusted). A "share split" is similar to a scrip/capitalisation issue, but is technically different.

The scrip issue (or share split) is likely to have been undertaken here because 900p per share is considered to be heavy, and a lower share price of 450p would help improve the marketability and liquidity of the shares. It can often be seen, for example, that shares with a lower price, such as 40p, tend to experience greater price changes in percentage terms (such as 40p to 44p: 10%) than is experienced with higher share prices (such as 400p). This is just one of many seemingly anomalous characteristics of the equity market.

The number of shares in issue will similarly change when, for example, a company is struggling, and its share price has fallen from 100p to 10p per share. One share could replace four shares, with the resultant share price being 40p.

Shareholders/investors, traders, analysts and financial commentators monitor such events closely. Anybody having a more passive interest in the share performance of a quoted company needs access to information that is sufficiently comprehensive and up to date to present a true picture.

When shareholders' entitlement to the dividend has just passed (and the share becomes "ex-dividend" — or xd), the share price will be adjusted downwards. "Cum-dividend" is when the shareholder will be entitled to the dividend payment. Similarly, "ex-rights" (xr), and "ex-capitalisation" (xc) will be applicable after a rights issue and capitalisation issue respectively. When share prices are listed in the daily newspapers and other media, xd, xr etc will be shown, although only for so long. "Ex-all" (xa) relates more broadly to whatever the company is issuing.

Net asset value

Another financial measure of a company is its "net asset value" — which is assets less liabilities (see chapter 7). This is more often used with property companies and investment trusts, for example, where asset values are closely linked to the share price — with share prices being referred to as having a "discount" or "premium" to net asset value.

Net asset value may also be considered where a takeover is being considered in order to establish, for example, if a buyer is paying £150m at a P/E ratio of 15 (based on £10m earnings), when the net asset value is £200m. The scurry for companies to appoint valuers when faced with a takeover bid was covered on page 27. Revaluations of property assets can also influence share prices.

Company accounts

A company's published accounts provide detailed financial information and also generally comment on the year's trading and future prospects. This includes the chairman's statement, directors' reports etc, which could be optimistic or pessimistic in tone — although sometimes, just misleadingly positive. Companies may announce profit warnings, or release other news that could adversely affect its future prospects and share price. Frequent changes of senior personnel, a change in auditors or qualifications in an auditor's report could be a sign of problems.

Directors' trading in the shares of their company are also monitored in the financial press, with sales, for example, suggesting a weak outlook and purchases a positive outlook. However, this is not necessarily reliable as directors may need money for private purposes, or an incoming director may buy shares in a struggling company as a sign of personal commitment. Sales/purchases by a number of directors tend to be more reliable, although could be a means of helping prop up a falling share price. There are also restrictions as to when directors can buy and sell shares in their own company.

The equity market and property

The fortunes of the equity market can affect the relative attractiveness of investment property as an asset class. Institutional investors, for example, may benefit from the greater stability offered by property, and its scope to secure both income and capital gains over a certain period of time. This scope for total return, such as (in simple terms) a 6% income yield and a 10% increase in capital value (total 16%) compares favourably, for example, with a fall in shares prices, as measured by the FTSE 100 index, from 6,000 to 4,500 (which represents a fall of 25%).

Increased demand for investment property decreases yields and increases capital/investment values. As shown in chapter 3, an increase in the capital value of investment property generally increases the viability of property development and therefore the level of development activity taking place. In contrast, a withdrawal of institutional investment from property will generally see yields increase, capital values fall and development become less viable.

In practice, there are many interrelated factors — relating to the economy and its various economic variables, the performance of the financial markets and the usual range of property issues that affect rental values and investment yields.

As examples of the indirect effects on the property market of the financial markets, a strongly performing stock market increases the level of trading activity, and therefore the demand for office space in the City of London financial core, which, in turn, increases rental values, with improved rental growth prospects also helping to lower yields. Salaries and bonuses in the financial markets influence the demand for typically higher-value residential properties, and this influences the value of residential properties in price brackets beneath.

At the lower end of the property investment market, capital values and yields are influenced particularly by interest rates and the cost of finance. Private investors, for example, see property as a more attractive opportunity than the low interest rates available on deposit accounts, and often judge the superiority of property by reference to the differential between investment yields and interest rates. Sometimes this is without due regard to the risk of yields increasing, capital values falling and losses being incurred. When investors need finance, they will be looking at the margins achievable between the cost of finance/interest repayments and the property yield achievable: the lower the cost of finance, the lower the yield and the higher the capital/investment value.

As another illustration of some of the apparently anomalous ways markets sometimes work, a fall in the value of equities while the outlook for property is good, could see institutional investors actually buy more equities and sell property. This is due to the need to maintain certain weightings between asset classes within a portfolio, and to ensure that investments remain within the necessary risk tolerances.

Other financial markets

The financial markets also include bonds/gilts, currencies/foreign exchange, commodities and a range of financial instruments that allow financial and trading risk to be managed. Financial instruments also enable investors to take highly geared (or "leveraged") positions, and trade against small price movements. "Derivatives" derive from the performance of a principal market and include options, futures and forwards contracts, and investment/trading can also take place against indices.

For a more comprehensive overview of the financial markets, publications are available in the larger bookshops — including *How to Read the Financial Pages* by Michael Brett (a regular contributor to *Estates Gazette* and author of EG Books' *Property and Money* (see *www.propertybooks.co.uk*), which provides a guide to commercial property investment and finance).

Accounting 7

This chapter explains the basic principles and components of company accounts, and shows how the financial strength of companies can be assessed. Importantly, it illustrates how accounting information can be a key driver in a business, as opposed to simply being a record upon which tax liabilities and other obligations are administered.

The first section in the chapter provides an example of a surveyor's accounts as a sole trader and shows how a basic profit and loss account and balance sheet are prepared, together with worked illustrations of accounting principles, including the balancing of accounts. The next section provides an example of a small limited company's "abbreviated" accounts and introduces further accounting entries.

A detailed example is then provided of the full accounts of a limited property investment company, explaining the various entries in relation to the company's trading activities, and providing initial illustrations of how accounts can be analysed. The next section illustrates the issues arising in the analysis of company accounts. Particular reference is made to retailing and manufacturing companies (as a professional services/surveying consultancy and property investment company are used for earlier illustrations). Property and company accounts is included as a final section, as prior coverage of other issues is necessary in order to put the issues into context.

The chapter concentrates on the analysis of small and mid-sized companies, as opposed to investors' and analysts' assessment of larger companies in connection with stock market investment, for example. Taxation is not covered in detail, but instead reference can be made to EG Books' separate publication, *Starting and Developing a Surveying Business: For Sole Traders* (see *www.propertybooks.co.uk*). This explains income tax, national insurance, corporation tax and VAT, as well as the differences between the trading status of a sole trader, partnership and limited company.

Once beyond the basics, surveyors will usually need to work closely with accountants. This is particularly important in view of the manipulation that can be made to company accounts and other financial information — and not just for tax efficiencies but to present a desired view to lenders, valuers, landlords, purchasers of a business or anyone else relying on the company's financial position to make decisions. Accounting practice and the treatment of certain items in accounts can also, legitimately, vary between companies, reflecting the nature of their respective businesses as well as accounting conventions and variations in terminology and presentational format. To accurately assess the current fortunes and future outlook of a company, knowledge is needed of the business itself, industry norms and such like. However, the key requirement for most surveyors is to be able to broadly judge a company's current and future prospects. This involves identifying the significant and sensitive items, and related commercial issues facing the business — rather than possessing a detailed working knowledge of accounting conventions. Assistance from accountants can then be sourced as appropriate.

Illustration 1 — the surveyor as sole trader

In order to show how a simple profit and loss account would be formulated, the following example is extracted from the separate EG Books' publication, *Starting and Developing a Surveying Business*. The surveyor has ended the first year of self-employment, working as a sole trader. VAT registration has not been necessary, so this is not added to the surveyor's fees/turnover, and expenditure is gross of VAT. There are no staff receiving a salary as the self-employed surveyor is able to undertake all administration personally. Home-working minimises other costs. A private car is available and the surveyor is able to charge the use of the car at a rate per mile.

Cash/funding

A business current account was opened and £15,000 deposited. During the year, the sole trader's drawings totalled £5000.

Income

Clients were invoiced for £32,000, with £29,500 having been received, and a further £2250 not having been invoiced in respect of completed work due for invoicing. The sole trader's work in progress is not included in the year's income. Total income, or "turnover", is therefore £34,250.

Expenses

Stationary was purchased at the outset for £575, and a subscription taken for *Estates Gazette* and *Property Week* totalling £300. Business cards and letterheads cost £300, and advertising £500. Internet and e-mail cost £200. An accountant was paid £350 for initial advice, and lawyers similarly paid £350. Professional indemnity insurance cost £600 and telephone charges amounted to £700. Business trips throughout the year involved hotel bills totalling £2217. Lunches, drinks etc amounted to £950 (although the tax rules do not allow such entertainment to be set against income/tax). Various information totalling £150 was ordered from RICS library and training/CPD events were attended, totalling £225. Maps and copyright etc totalled £200, and a further £500 was incurred on a range of printing jobs. Stamps were purchased on a number of occasions and totalled £165. An RICS subscription of £390 was paid. In addition to a variety of petty cash expenses totalling £150, partial records of non-receipted photocopying costs, and journals are available, with the total estimated at £250. In addition to the actual items of expenditure, £300 is to be claimed as an expense in respect of heat and light for home-working. Mileage costs, calculated in accordance with the tax rules, total £4669. Total expenses amount to £14,041.

Capital items

Computer equipment, camera etc for the business was purchased at the outset for £3000, and office furniture for £1000. A percentage of the cost of capital items, known as "capital allowances", is set against profit. Whereas stationery, for example, is an expense set against income in a particular year, computer equipment or furniture are capital items set against income over a number of years. In accordance with the tax rules for small businesses, the computer and camera, at £3000, can be written down at 100%, and the furniture, at £1000, can be written down in the first year at 40% (and 25% per year thereafter). The capital allowances for this year are therefore £3400 (£3000 and £400).

In practice, a sole trader would complete the Inland Revenue's self-assessment forms as part of the year-end tax return (see *Starting and Developing a Surveying Business*). A basic accounting summary is shown, which is, in effect, a profit and loss account.

Table 7.1 Profit and loss account

	(£)
Turnover	34,250
Expenses	14,041
	20,209
Add disallowable expenses	£950
	21,159
Less capital allowances	3,400
Profit	17,759

The above illustrations are for taxation purposes, and this is not necessarily the same as the actual profitability of the business. Disallowable expenses, for example, are a business cost, influencing the actual profitability of the business (and in the accounting summary above, disallowable expenses are included in expenses, but then added back — rather than simply ignored). As another example, capital equipment will have a life longer than the one year assumed by the 100% writedown. The surveyor/company may therefore prepare different accounts for non-tax purposes that show a more realistic measure of profit, with disallowable expenses being accounted for, and capital equipment being written off/depreciated over a longer period of time. Page 135 includes further comment on depreciation.

A balance sheet prepared at the end of the year is shown in table 7.2, in line with headings on the Inland Revenue's self-assessment forms (which although not all applicable to professional/surveying services, would be to other businesses).

Table 7.2 Balance sheet

	(£)
Assets	
Plant and machinery	600
Other fixed assets	–
Stock and work in progress	–
Debtors/prepayments/other	4,750
Bank/building society balances	26,428
Cash in hand	–
	31,778
Liabilities	
Trade creditors/accruals	4,969
Loans and overdrawn bank accounts	–
Other liabilities	–
	4,969
Net business assets	**26,809**
Represented by (capital account)	
Balance at start of period	15,000
Net profit/loss	16,809
Capital introduced	–
Drawings	(5,000)
	26,809

To understand the concept of accounts, it is helpful to think that everything has to balance, and that there must be two elements to each transaction taking place.

- Plant and machinery is net of capital allowances for the year (example of balancing: the reduction in assets is a reduction in profit, with the bank account being unaffected as expenditure is not actually incurred).

- Debtors/prepayments/other is the fee income not yet received but included in the income for the year (example of balancing: income is added to profits despite not being received, so needs to be added to debtors/prepayments/other).
- Trade creditors/accruals is home/heat and light and mileage that are yet to be paid from the business account to the surveyor's personal account (example of balancing: expenses incurred are matched either by a deduction from the bank account or an addition to trade creditors/accruals).
- The net business profit for tax purposes, upon which income tax and national insurance for the year will be based, is £17,759, but for the balance sheet needs to be £16,809 as this represents the actual profitability of the business — with £950 being disallowable expenses for tax purposes. Income tax and national insurance is not an amount owing as a trade creditor/accrual, as this is the liability of the surveyor, not the business. However, if this were a limited company, corporation tax would be the liability for the business and included as trade creditors/accruals (see p108).
- The bank/building society balance can be checked by bank balance at start (£15,000), add income received (£29,500), less drawings (£5000), less capital expenditure (£4000), less expenses incurred/amount actually paid — ie excluding home expenses and mileage (£9072) = £26,428.

Bank/building society balances should correspond with the balance on the bank statement. If there was a cash balance, such as because an amount was withdrawn for petty cash, there would be both bank/building society balances and cash-in-hand entries. The check on the cash-in-hand balance is the actual amount of cash spent.

Illustration 2 — small limited companies

The accounts of a small limited company are publicly available from Companies House, but only limited information has to be provided by way of "abbreviated accounts". This generally makes it impossible for others to gauge profitability (although if profitability is high and profits/dividends not drawn, the increase in assets/shareholders' funds could give an approximate indication of the scale and success of the business).

In the following example, the surveyor's income/turnover is £80,000, which includes £10,000 not yet received (ie debtors). Expenses are £25,000, including £3000 not yet paid (ie creditors). The surveyor would be VAT registered as turnover is above the level for compulsory registration, and the above amounts for income and expenses are net of VAT. In practice, any VAT received on income or paid on expenses, but not yet accounted to Customs & Excise, would be creditors if a net amount were owing, or debtors if a net reimbursement was due. The surveyor does not withdraw a salary, but instead takes out dividends of £20,000. The "issued" share capital of the company is assumed to be £1000, which represents the initial amount committed by the surveyor for 1000 shares at £1 nominal/par value each.

The actual profitability of the business can be easily calculated at £55,000 (£80,000 turnover less £25,000 expenses). Corporation tax is payable at the small companies' rate of 19%, equating to a tax liability of £10,450 (and as not having been paid to the Inland Revenue at the end of the company's financial year is a creditor, taking total creditors to £13,450). Dividends of £20,000 would have been paid to the surveyor prior to the year-end, and leave a retained profit of £24,550 (£55,000 less £10,450 corporation tax, less £20,000 dividends).

An abbreviated balance sheet would be on the lines of the following:

Table 7.3 Abbreviated balance sheet

	(£)
Current assets	
Debtors	10,000
Cash at bank and in hand	29,000
	39,000
Creditors	
Amounts falling due within one year	13,450
Net current assets	25,550
Total assets less current liabilities	**25,550**
Capital and reserves	
Called-up share capital	1,000
Profit and loss account	24,550
Shareholders' funds	**25,550**

A check on the accounts can be made by examining the cash balance: initial cash balance of £1000, add £70,000 income received (net of VAT), less £22,000 expenses paid (net of VAT), less £20,000 dividends = £24,550.

Had a salary been paid to the director, this would be an expense that reduced profit and therefore corporation tax liability — with the surveyor having personal income tax and national insurance liability in the usual way. *Starting and Developing a Surveying Business* includes worked examples in respect of taxation, as well as illustrations of the tax efficiencies achievable by small surveying businesses.

In the above accounts, "net current assets" and "total assets less current liabilities" are the same, but if there were longer-term liabilities, such as outstanding loans, this would appear in the accounts between these two entries. Total assets less current liabilities is also called "net assets", and this is the same as "shareholders' funds". This represents the value of the business to the shareholders/owners (although see comments later in the chapter on goodwill and other factors).

In the example, full accounts would be required for the Inland Revenue. Further examples of accounts are shown within the investment company illustration in the following section, and at the start of Analysing company accounts on p126.

Other issues

In order to explain other accounting entries before examining the investment company's accounts, fixed assets comprise tangible fixed assets, such as property, plant and machinery, vehicles, fixtures and fittings; and intangible fixed assets, such as goodwill (see p134) and patents.

Current assets include debtors, prepayments, cash/bank account deposits and stocks. Stocks include raw materials, work in progress and finished goods, and are included in the accounts at the lower of their cost, or net realisable value. Cost of work in progress and finished goods includes the actual cost of the goods, labour and a proportion of overheads, which do not go to the profit and loss account. Net realisable value is the expected proceeds of sale less the expected selling costs.

Current liabilities include bank overdrafts, creditors, accruals, hire purchase and other finance. Creditors, as part of current assets, generally include amounts owing in less than one year. Creditors also includes corporation tax due and dividends that have been declared but not paid. As mentioned above, VAT owing to Customs & Excise would be a creditor (and a debtor if amounts were due from Customs & Excise).

Capital and reserves include share capital and reserves. Share capital is issued share capital, and comprises the nominal/par value of the issued shares (£1 in the above example of the limited company, which is not the same as a company's share price). "Share premium account" sometimes appears in the accounts, and reflects the issue of new shares above their nominal/par value, and therefore enters the "premium" element as share premium (which is the premium above the nominal/par value).

The format and content of company accounts, including "disclosure" requirements, are governed by company law and accountancy standards. These are designed to provide consistency between financial statements, and generally ensure that a "true and fair view" is given of companies' trading and financial affairs. Examples include the Companies Act 1985, Financial Reporting Standards (FRS), Financial Reporting Standards for Smaller Entities (FRSSE), General Accepted Accounting Practice (GAAP), Statements of Standard Accounting Practice (SSAP), Urgent Issues Task Force (UITF) Abstracts, International Financial Reporting Standards (IFRS) and International Accounting Standards (IAS). The Accounting Standards Board (ASB) issues accounting standards in the UK, and the International Accounting Standards Board (IASB) works on standards on a worldwide basis. Countries can decide whether to implement international standards, although in the case of the UK, for example, standards could be imposed by company law. This chapter does not reference the accounting principles to such rules as surveyors will rarely need to be familiar with such detail.

Illustration 3 — investment company

This section illustrates the accounts of a small plc property investment company, and also the surrounding business issues relevant to the surveyor. Comment on general accounting matters is included.

Nature of the business

The company is a rapidly growing family business that has been heavily reliant on the support of lenders, and has benefited from successful letting and management functions. In managing the business around often squeezed cash flow, and the need to still present a favourable, relatively low-risk view to risk-conscious lenders and others, property decisions are linked closely to accounting and financial matters. Approximately 70% of the portfolio, by rental, is residential, and 30% is commercial. In addition to some well-let commercial investments and professional residential lets, the portfolio comprises multi-let houses, including to students. Overall, the portfolio is high yielding, with active management being concentrated into a geographical area of approximately two square miles in a major city (although there are a small number of properties beyond).

Surveyors' business skills input

The illustration is particularly relevant to *Business Skills for General Practice Surveyors* as it shows how, with the right business skills, a surveyor can be a driving influence in the development of client businesses. This includes the development of the right profile of a client's business in terms of its marketing activities.

General information

The cover page of the investment company's accounts is headed "Report and Financial Statements for the year-ended xxxx", together with the company name, "Registered in England and Wales" statement and the name and address of the accountants. The contents page lists company information, report of the directors, report of the auditors, profit and loss account, statement of recognised gains and losses, balance sheet, cash flow statement and notes to the financial statements. "Company information" in the accounts includes details of directors and the secretary, the registered office, the registered number, the date of incorporation, the auditor (which is the same as the accountants) and the bankers/lenders.

Report of the directors

The "report of the directors" begins with a statement of "The directors present their report and accounts and the audited financial statements for the year-ended xxxx". Under a heading of "Principal activities", it then states that "The principal activities of the company are that of purchasing properties for the purpose of letting". The report and accounts are completed three months after the year-end, and reference to the "current year" below means the year after the recent year-end.

Under "Business review and future developments" within the report of the directors, the following commentary is provided by the managing director, who is a majority family shareholder working in a full-time capacity for the business, and has a building contracting background rather than surveying and investment expertise.

There has been a significant increase in the number of properties purchased during the year, some of which were properties that had been let or were already let but in a poor state of maintenance. In some properties, tenants were moved, leaving the properties to be completely refurbished. Refurbishments were continuing up to xxxx (three months after the year-end, and to the date of the report). In most of the properties that were acquired, furniture and fittings had to be replaced. Due to the timing of delays in carrying out refurbishments, the accounts do not reflect the full year's income on all of the properties. It is anticipated that if all of the properties are let, then a full year's income would be approximately £590,000. This will not be achievable during the year to xxxx (end next year) due to refurbishments continuing into the current year. In any event, this will never be 100% achievable due to void periods and the changeover of tenants during each letting year.

It is anticipated that the forthcoming year will be one of consolidation, with few properties being purchased. During this period of consolidation, the directors intend to carry out a maintenance programme in order to keep the workforce occupied.

In order to consolidate and simplify the company's accounting procedures, the properties were revalued by xxxx (a large international practice) on behalf of the company's main mortgage providers, namely xxxx and xxxx. The help and support of these two mortgage providers is greatly appreciated.

It is with regret that xxxx, chartered surveyor, appointed during the year, does not seek re-election as a director of the company due to his other professional commitments. xxxx is available to undertake professional assignments as and when necessary. It is xxxx's input during the last year that is largely responsible for the dramatic increase in the growth of the portfolio of properties.

This overview on the trading year and its future prospects is an important way of putting the financial performance into context. As is demonstrated below, the company actually made a loss in the year. However, closer examination of the accounts, and more knowledge of the "growth story" and the nature of the business (including the property opportunities it exploits) show the company to be inherently very profitable.

The report of the directors then lists the directors. Directors' shareholdings are listed at the end of the final year to which the accounts are prepared, and for the previous year-end. "Family shareholdings" are also stated (as the wife of one of the directors, who was company secretary, was not actually a director). Other statements include confirmation that the company made no charitable donations. A statement is also provided that "turnover and operating profit derive wholly from continuing operations". Accounts should also highlight whether there are any discontinued activities reflected in the financial statements, as these would, of course, influence interpretation of the accounts and comparison on a year-to-year basis. As a general point, various other statements will be seen, such as exemption from audit and references to sections of the Companies Act 1985.

Auditor's report

Under a heading of "Auditor's report to the shareholders of xxxx", auditors provide the statement set out below. Although relatively standard, it shows an auditor's role and responsibilities.

We have prepared the financial statements on pages xx to xx, which have been prepared under the accounting policies set out on pages xx and xx.

Respective responsibilities of directors and auditors
As described on page xx, the company's directors are responsible for the preparation of the financial statements. It is our responsibility to form an independent opinion, based on audit, on those statements and report to you.

Basis of opinion
We conducted our audit in accordance with Auditing Standards, issues by the Auditing Practices Board. An audit includes examination on a test basis, of evidence relevant to the amounts and disclosures in the financial statements. It also includes an assessment of the significant estimates and judgments made by the directors in the preparation of the financial statements, and of whether the accounting policies are appropriate to the company's circumstances, consistently applied and adequately disclosed.
We planned and performed our audit so as to obtain all the information and explanations that we considered necessary in order to provide us with sufficient evidence to give reasonable assurance that the financial statements are free from material misstatement, whether caused by fraud or other irregularity or error. In forming our opinion, we also evaluated the overall adequacy of the presentation of information in the financial statements.

Opinion
In our opinion, the financial statements give a true and fair view of the state of the company's affairs as at xxxx (year-end) and of its loss for the year then ended, and have been properly prepared in accordance with the provisions of the Companies Act 1985.

As a general point, although an accountant can prepare the accounts as well as do the audit, they will not necessarily be familiar with the day-to-day running of the business. High-profile accountancy scandals, and misdemeanours committed by smaller companies, can go unnoticed by auditors. Therefore, it is important that directors and surveyors, as advisers, are alert to the potential indiscretions of others.

Set out initially below are the financial statements included in the accounts. More detailed comments are then provided, including with reference to "notes to the financial statements" (with further more general accounting concepts being explained later under the section "Analysing company accounts"). The accounts list figures for the year to which the accounts relate, and also the previous year (and redated years of 2003 and 2002 are adopted below). Where brackets are used, this denotes losses or a deduction (as accounts will not say "add" or "less", for example, and users should generally know whether entries are being added or deducted). The accounts are also signed by one of the directors, with a statement to the effect that the financial statements were approved by the board. The headings in the accounts, as with the various inclusions mentioned above, reflect statutory requirements.

Table 7.4 Profit and loss account

	2003 (£)	2002 (£)
Turnover	475,196	391,337
Staff costs	107,347	75,792
Depreciation and amortisation	17,688	15,041
Other operating charges	268,183	138,097
	(393,218)	(228,930)
Operating profit	81,978	162,407
Interest (payable)	(103,092)	(87,993)
(Loss)/profit on ordinary activities before taxation	(21,114)	74,414
Taxation	2,641	(17,625)
(Loss)/profit for the year	(18,473)	(56,789)
Retained profit brought forward from previous years	139,574	82,785
Bonus share issue	(84,627)	–
Retained profit carried forward	36,474	139,574

Table 7.5 Statement of total recognised gains and losses

	2003 (£)	2002 (£)
(Loss)/profit for the financial year	(18,473)	56,789
Increase on revaluation	677,970	–
	659,497	56,789

Table 7.6 Reconciliation of movements in shareholders' funds

	2003 (£)	2002 (£)
Total recognised gains and losses for the year	659,497	56,789
Shares issued for cash	62,500	–
Shareholders' funds at (end previous year)	814,477	757,688
Shareholders' funds at (end current year)	1,536,474	814,477

Table 7.7 Balance sheet

	2003 (£)	2002 (£)
Fixed assets		
Intangible – Goodwill	15,000	20,000
Tangible – Investment properties	3,413,000	1,778,500
– Other fixed assets	107,527	24,950
	3,535,527	1,823,450
Current assets		
Debtors due within one year	119,232	7,292
Cash at bank	28,889	3,415
	148,121	10,707
Creditors		
Amounts falling due within one year	(1,007,156)	(544,735)
Net current assets (liabilities)	(859,035)	(534,028)
Total assets less current liabilities	2,676,492	1,289,422
Creditors		
Amounts falling due after more than one year	(1,133,018)	(474,945)
Provisions for liabilities and charges		
Deferred taxation	(7,000)	(–)
TOTAL NET ASSETS	**1,536,474**	**814,477**
Capital and reserves		
Called-up share capital	1,500,000	500,000
Revaluation reserve	–	174,903
Profit and loss account	36,474	139,574
SHAREHOLDERS' FUNDS	**1,536,474**	**814,477**

Cash flow statement

The cash flow statement reconciles cash flow with movements in bank balances, debtors and creditors. It ignores depreciation and movements in tax provisions and reserves.

Table 7.8 Cash flow statement

	2003 (£)	2002 (£)
Net cash flow from operating activities	28,169	182,340
Returns on investments and servicing of finance:		
Interest paid	(103,092)	(87,993)
Taxation	(17,484)	(25,236)
Capital expenditure and financial investment	(1,056,395)	(4,225)
Cash (outflow)/inflow before financing	(1,148,802)	64,886
Financing	1,183,169	2,464
Increase in cash in the period	34,367	67,350
Reconciliation of net cash flow to movement in net debt		
Increase in cash in the period	34,367	67,350
Cash flow from increase on debt	(1,120,669)	(2,464)
Change in net debt resulting from cash flows	(1,086,302)	64,886
Net debt brought forward	(957,039)	(1,021,925)
Net debt carried forward	(2,043,341)	(957,039)

Notes to the financial statements

The notes to the financial statements begin with the following statement under "Directors' responsibilities", which illustrates the duties placed on directors and the basis on which company accounts are prepared.

Company law requires the directors to prepare financial statements for each financial year that give a true and fair view of the state of the company and of the profit and loss of the company for that period. In preparing these statements, the directors are required to:

- Select suitable accounting policies and then apply them consistently.
- Make judgments and estimates that are reasonable and prudent.
- Prepare the statements on a going concern basis unless it is inappropriate to presume that the company will continue in business.

The directors are responsible for keeping proper accounting records that disclose with reasonable accuracy at any time, the financial position of the company, and to enable them to ensure the financial statements comply with the Companies Act 1985. They are also responsible for safeguarding the assets of the company and hence for taking reasonable steps for the prevention and detection of fraud and other irregularities.

The notes to the financial statements also comment on the accounting policies adopted. Companies are required to select appropriate accounting policies and accounts must also be prepared in line with accounting standards.

The notes to the financial statements also provide more detailed explanations of some of the entries in the accounts provided above — such as interest payable (£103,092) being separated into "bank overdraft and loan interest" (£48,342), "mortgage interest" (£54,620) and a small amount of "interest on overdue taxation" (£130). Loan repayments are broken down into "less than one year or on demand", "between one and two years", "between two and five years" and "after five years".

Breakdowns are provided of:

- directors' remuneration (salaries and pension contributions)
- staff costs (wages and salaries, social security costs and other pension costs)
- investment properties (fixed assets)
- other fixed assets and related depreciation (motor vehicles, office fixtures and fittings for the company's offices and fixtures and fittings included in the investment properties)
- goodwill.

Breakdowns are also shown of debtors, creditors and borrowings. Details are provided on share capital, the property revaluation reserve and the issue of bonus shares (see later). More detailed information on profit, cash flow and debt analysis is also provided.

Additional information

Additional information is included in the full set of accounts but this does not form part of the statutory accounts — accompanied by the statement, "The following pages do not form part of the audited financial statements and are for management purposes only". This enables further detailed information to be presented to lenders, shareholders, potential new investors etc, but which need not be made available more widely (ie filed at Companies House for public access by creditors, competitors or others).

As the company was looking to grow quickly, mainly through reliance on finance, lenders' support was imperative. As indicated above, it was important to maintain the right impressions, particularly in respect of risk profile, calibre of management and investment/property management expertise. Three-monthly "management accounts" were therefore prepared (comprising a full profit and loss account) complemented by reports and correspondence providing updates on the company's activities.

Such detailed, up-to-date accounting information can be achieved relatively easily through well-organised administration and management systems, including the computerised use of accounting formats. This supports key decision making in respect of investment strategy, including the affordability of further acquisitions, the type of properties that could be acquired, and the likely arrangements in respect of finance. The investment company's largest returns, for example, would derive from refurbished/repaired properties, but these involve works expenditure and no immediate income/rent, while interest repayments to lenders have to be made (ie in contrast to the standing investment property that immediately provides a rent). The ability for the company to undertake the work cost-effectively with a mainly in-house workforce, is a good example of how investment activity combines with development/refurbishment activity to help maximise profits. The cash flow requirements this creates further reiterates the importance of finance and in securing support from lenders: the importance of the profile of the company, the quality of accounting information maintained, and a track record in managing a precarious cash flow position and any other risk factors.

Additional profit and loss information

The additional information includes a more detailed profit and account, which is as follows:

Table 7.9 Profit and loss account in more detail

	2003 (£)	2002 (£)
Rents receivable	475,196	391,337
Deduct overheads and expenses		
General repairs	160,676	75,820
Secretarial salary	25,600	24,300
Office salary	13,375	10,050
Wages	24,546	2,500
National Insurance	9,259	6,636
Rates and water	10,098	7,401
Light and heat	16,979	20,946
Insurance	8,968	6,478
Transport	8,675	5,287
Printing, stationery, advertising	5,352	269
Bank charges and interest	55,081	48,992
Mortgage interest	54,620	39,405
Professional fees	18,481	8,991
Telephone charges	3,056	2,420
Sundry trade expenses	1,321	586
Audit and accountancy charges	23,879	8,802
Interest on overdue taxation	130	693
Theft of materials	3,959	–
	(444,055)	(269,576)
Net profit before taxation, depreciation and directors' remuneration	31,141	121,761

	2003 (£)	2002 (£)
Depreciation and amortisation	(17,688)	(15,041)
Directors' remuneration	31,900	30,600
Contributions to directors' pension scheme	2,667	1,706
	(34,567)	(32,306)
Net (loss)/profit before taxation	(21,114)	74,414
Taxation	2,641	(17,625)
(Loss)/profit for the year	(18,473)	(56,789)
Retained profit brought forward from previous years	139,574	82,785
Bonus share issue	(84,627)	–
Retained profit carried forward	36,474	139,574

The "net (loss)/profit before taxation" entry (£21,114 loss in 2003 and £74,414 profit in 2002) correlates with the entry in the profit and loss accounts in the statutory accounts (p113) of "(loss)/profit on ordinary activities before taxation", but the ordering of other elements is different.

The significant factors from the cost side of the profit and loss account are highlighted as:

- Repairs are approximately £160,000.
- Interest repayments/banking charges total £109,831. (This is a different figure to the interest payable figure of £103,092 in the profit and loss account on p113, because bank charges would have been included in "other operating charges" in the profit and loss account.)
- Entries of rates and water, light and heat and insurance, total approximately £30,000. As the company has only a small office for itself, these clearly relate to lettings where rents are inclusive of such items (noting also that the company is responsible for the insurance of all properties).

The arrangements in respect of remuneration of directors and professional advisers would also need examining, although with much analysis of company accounts, the emphasis on certain areas of analysis depends on the purpose of the analysis. Scrutiny of the basis on which fixed assets are entered in the balance sheet would be made, together with the extent of borrowing (or "gearing") and the solvency of the company. The company has made a loss for the year, which would typically be a cause for concern.

The above factors are drawn on in subsequent analysis below.

Year-on-year trends

One year's accounts is not sufficient to be able to make accurate judgments on a company's fortunes and future prospects. Three or more years' accounts are ideally available although, in the case of new companies, information will be limited.

The additional information for the investment company which did not form part of the statutory accounts, also included a "summary of profit and loss accounts for the five years ended 2003", and a summary of balance sheets for the same period. For any business, this shows year-on-year trends, which, as mentioned under the section "Analysing company accounts" later, is important. For the investment company concerned, it also shows the growth record, helping demonstrate the success and future potential of the business to lenders, new shareholders etc. The management of the company's growth plans was, in fact, pursuant to an option of stock market flotation several years ahead, and this is an another example of how accounting and financial information, property decisions and growth strategy need to reflect such business/clients' objectives.

Extracts from the accounts show "net (loss)/profit before taxation" as follows:

Table 7.10 Profitability over five years

2003 (£)	2002 (£)	2001 (£)	2000 (£)	1999 (£)
(21,114)	74,414	71,645	112,373	57,322

Fixed assets are shown as follows (which predominantly represent investment property — ie £3.4m property out of £3.5m total assets in the last year, as shown in the balance sheet on p114):

Table 7.11 Fixed assets/property over five years

2003 (£)	2002 (£)	2001 (£)	2000 (£)	1999 (£)
3,535,527	1,823,450	1,834,266	1,026,401	587,318

Losses and full potential

This shows that five years ago profits of approximately £60,000 were made on fixed assets/investment property value of approximately £600,000 (ie 10%) and the next year, profits of £110,000 were made on assets/investment property of £1m (ie 11%). In contrast, in the last financial year, a loss of approximately £20,000 has been made on assets/investment property value of £3.5m. However, growth has been strong, with an approximate sixfold increase in fixed assets/investment property being experienced over five years — achieved primarily by further property acquisitions, although partly also by property revaluations (see later).

As indicated in the report of the directors, losses reflect the considerable number of acquisitions taking place in the year, particularly of properties in need of repair/refurbishment. This is suspected by the entry of repairs in the profit and loss account, but this is not a conclusive indication owing to the accounting treatment of repairs/refurbishment (see later). Such properties are incurring expenditure on works, and expenditure on interest repayments, while not receiving any rental income/turnover. Therefore, although such properties ultimately provide high returns, the short-term financial position appears problematic. The accounts also show that investment properties, as part of fixed assets, represented £1,778,500 at the start of the year, and that £956,530 additions were made at cost, with revaluations amounting to £677,970 (and totalling the £3,413,000 entry in the balance sheet on p114). At cost, the portfolio had therefore grown during the year by 54%, and by total value had increased by 92%. Additions at cost include refurbishment expenditure capitalised to the balance sheet, as well as acquisitions (see explanation later).

As an illustration to show how an apparently problematic financial position works its way out over time: if the company ceased acquiring new properties, and just managed a fixed portfolio (undertaking only essential maintenance and reducing other costs), the business would be securing approximately £570,000 income (after 5% voids) on £3.75m fixed assets/property values (say £0.25m higher than £3.5m as refurbishment/repair would add to the fixed assets/property values). After deducting buildings insurance, rates and water, and heat and light paid for by the company in order to achieve a net property rent (more equivalent to a full repairing and insuring (FRI) basis), income would be around £525,000. Against £3.75m, this reflects a yield of 14%, reflecting the nature of the portfolio and its active management. Rents are also relatively high, reflecting the superior internal condition and facilities of the company's residential properties compared with other landlords, and the strong brand created by the company in the market — ie the company is profiled suitably for lenders, shareholders, new investors etc, and also for its customers/clients/tenants. Commercial properties are also high yielding, reflecting their secondary location and pro-active management strategy. After interest repayments of approximately £175,000, and other costs of around £75,000 (which reduce considerably if expansion is not being pursued) profit before directors'/shareholders' remuneration is some £275,000. This is a good illustration of an active property investment business. In comparing £275,000 profit before directors'/shareholders' remuneration to, say, 30% of £3.75m property values (£1.125m) to represent gearing of approximately 70% (as a loan-to-value ratio), directors'/shareholders' return on equity is approximately 25% pa. These are just illustrative calculations, with more formal measures of profitability being considered in the section "Analysing company accounts".

Fixed assets, revaluations, repair/refurbishment

As indicated within the comments above, other aspects for scrutiny include the basis on which property valuations (including revaluations) are included in the accounts above, and how repair/refurbishment expenditure is treated.

The accounts stated that:

> Properties with a total value of £2,908,000 were valued by xxxx (international property consultants) in xxxx (two months prior to the year-end) at market value. The directors consider that these values are appropriate at the year-end. The remaining properties with a total value of £505,000 were valued at the year-end by the directors on the basis of bank/building society valuations for mortgage purposes obtained during the year.

The valuation by the international property consultants was part of a major remortgaging exercise, and valuations specifically for company accounts purposes were not needed. Although the directors were able to enter their own opinion of valuations, and this may suggest scope for impropriety, a proper approach was imperative — for reasons including the need to create the right impression to lenders, and to avoid subsequent valuations by appointed valuers differing considerably. As with the overall accounts, the accurate representation of property values was important in view of the scope to raise capital from existing and any new investors/shareholders. It was also important because the amounts committed would be pegged to the value of the company/shareholders' funds per share — in effect the right share price.

Also, although the company was pursuing growth rapidly — drawing on finance and operating to a sometimes precarious cash position — risk also had to be managed. Examples of the prudent approach to risk included the acceptance of cautious valuations (rather than overvaluation by keen to please valuers), a mix of variable and fixed rate finance, loans on an interest-only basis as well as those involving loan repayments, holding a secured overdraft on one property, and directors, family and certain other shareholders having a proportion of sufficiently liquid funds as part of their own personal financial arrangements that could be committed if need be.

Notwithstanding the accuracy of the investment company's approach to fixed assets, there are a number of issues that surveyors need to consider in connection with the representation of property in company accounts and regarding taxation issues. In the case of the investment company, a major issue is the treatment of refurbishment and repair expenditure. In simple terms, general maintenance and repair would be a cost in the profit and loss account, thus reducing profit and also taxation liability for the year. More extensive refurbishment would effectively be capital expenditure and would need "capitalising", ie would be added to fixed assets, appearing as such in the balance sheet and, although not being set against profits, would be included in any capital gains tax computation on sale. Repair costs set against taxation was preferable for the company and, in view of the importance of observing the relevant tax rules, close liaison had taken place with the Inland Revenue on the approach to be taken. Tax savings, as with any cost savings, are effectively a source of funding. This is a good example of where surveyors need to take detailed accountancy and/or legal advice in respect of their surveying work — the business skill being the recognition of such issues, rather than detailed knowledge. Capital allowances were not significant to the subject investment business, but present a source of savings/gains in other situations.

Accounts examination — profitability and remuneration

Within company accounts, it is important to examine how directors' remuneration has been treated, and how this affects actual profitability and measures of profitability. Directors' salaries will be a cost that is set against turnover and reduces profit, and this is why it was helpful to see directors' remuneration expressly accounted for in the additional information to the profit and loss account. Directors may not necessarily be shareholders, in which case remuneration will comprise only their salary and any benefits. If they are shareholders, earnings potentially comprise a combination directors' salary and benefits, dividend payments and an increase in the value of the shares held in the company (and possibly share options or free shares). In the case of the above company, directors were also shareholders, which is why directors' salaries were low — and really only to cover the living expenses of a single, full-time/managing director. The director's wife was company secretary, and a shareholder, but not a director, with her secretarial salary appearing alongside the main costs. The "office" salary was also in respect of a family member. In other situations, relatively low remuneration in relation to directors' positions/status and typical remuneration in similar companies could indicate that the company is struggling, notwithstanding an apparently satisfactory level of reported profit.

These finer issues are important when considering the actual profitability of a business, bearing in mind also the purpose for which such assessment is being made. Purposes include a bank's lending decisions, how much a prospective new investor pays for shares, the viability of the business to its principals/shareholders/family against other business and lifestyle options available, the price to be paid by a purchaser for the business, or similarly the value/share price on stock market flotation. As an example, relatively low remuneration to principals/shareholders/family by way of salaries in the profit and loss account would flatter profitability. Instead, remuneration can be provided via dividends and an increase in the value of shareholdings. An individual or company taking over the business and relying on new salaried staff who were not shareholders would instantly incur increased costs and suffer lower profitability. The pros and cons of the forms of remuneration for principals/shareholders/family do, of course, also depend on whether there are other shareholders, and whether they take an active role in the business. With the above company, there were other family shareholders who had committed capital to help the business expand, but were not involved in the day-to-day affairs. The chartered surveyor referred to in the report of the directors also stood as a director during the year and additionally held shares in the company as part of overall arrangements.

Share capital

As stated on p114 in the balance sheet, total net assets and shareholders' funds were:

Table 7.12 Balance sheet extract

	2003 (£)	2002 (£)
TOTAL NET ASSETS	**1,536,474**	**814,477**
Capital and reserves		
Called-up share capital	1,500,000	500,000
Revaluation reserve	–	174,903
Profit and loss account	36,474	139,574
SHAREHOLDERS' FUNDS	**1,536,474**	**814,477**

Details of the share capital of the company were in the notes to the accounts as follows:

Table 7.13 Notes to the accounts — share capital

	2003 (£)	2002 (£)
Authorised		
2,000,000 (2002: 1,000,000) ordinary shares of £1 each	2,000,000	1,000,000
Called-up, allotted and fully paid		
500,000 ordinary shares of £1 each	500,000	500,000
62,500 ordinary shares of £1 each issued for cash	62,500	–
937,500 bonus issue shares	937,500	–
	1,500,000	500,000

The issue of bonus shares involves capitalising reserves into shares. The £937,500 issue of bonus shares is facilitated by £852,873 from the revaluation reserve (£174,903 at the start of the year, £677,970 revaluations during the year, and £nil at the end of the year) and £84,627 from the profit and loss account (£139,574 at the start of the year, less post-tax loss of £18,473 for the year, less £36,474 retained profit), totalling £937,500.

The issue of bonus shares for the investment company is facilitated by the increase in the "revaluation reserve" because of property revaluations. If properties are revalued at a higher amount than already listed as fixed assets, the increase is represented in the balance sheet by both an increase in the fixed assets entry and a corresponding entry to "revaluation reserve". The issue of 62,500 ordinary shares was for cash. This was at the nominal/par value of £1 per share, and reflected the value of the shares considered to be £1 each (which based on shareholders' funds of £1,536,474 ÷ 1,500,000 shares following the bonus issue is conveniently accurate and is achieved by the amount of shares/reserves used for the issue). If 62,500 new shares were issued at a higher price than the nominal/par value, at, say, £1.20 each (£75,000), the balance sheet would still include called-up share capital at £1,500,000 (including the new shares at £62,500) but would also include £12,500 in a "share premium account" (£75,000 – £62,500, or 20p per share premium). The issue of new shares also increases the cash/ current assets balance, and therefore total net assets, with the revised section of the balance sheet being:

Table 7.14 Share capital — revised illlustration

TOTAL NET ASSETS	**1,548,974**	**814,477**
Capital and reserves		
Called-up share capital	1,500,000	500,000
Share premium account	12,500	–
Revaluation reserve	–	174,903
Profit and loss account	36,474	139,477
SHAREHOLDERS' FUNDS	**1,548,974**	**814,477**

The issue of new share capital from reserves (such as the profit and loss account/ reserve, the revaluation reserve or the share premium account) locks in the capital, which is particularly beneficial in the eyes of the investment company's lenders and any shareholders/investors not actively involved in the business as directors (and likewise creditors in other situations). The capital is locked in because dividends can only be distributed from "distributable reserves", which is the accumulated/retained profit in the profit and loss account/reserve (and any other reserves emanating from accumulated/retained profit, which may have been separately named). Dividends cannot be distributed from "undistributable reserves", which include the revaluation reserve (as any profits have not been realised), the share premium account or a capital redemption reserve (which is beyond the scope of this chapter). Other restrictions are placed on the reduction in capital (such as a company purchasing shares from shareholders), and in the case of plc companies, the distribution of dividends cannot be made if it would cause the net assets figure to fall below called-up share capital. Also, in the case of a company trading property (ie buying to sell in the short term rather than longer-term investing), a deferred tax reserve may be used to account for the likely tax liability on the sale of assets (with the asset value being carried in fixed assets and the revaluation reserve net of such liability).

An example of some of the finer business issues is that an issue of bonus shares to shareholders/investors can simply be judged by them as being good, and reiterate the success of the venture — even though total shareholders' funds and the value of shareholdings do not change.

Surveyors and business skills

It can be seen from the example that the company's trading year has involved an aggressive acquisition programme, making the absolute most of cash that is available and, at the same time, making prudent acquisitions that provide the best returns, and fit the existing portfolio and overall business plan. In order for the surveyor to help take the business forward, an understanding is needed of company accounts, and their impact on lenders, investors etc. This is just one illustration of how business skills enhance a surveyor's ability to perform on behalf of clients, and also share in the due financial rewards. In practice, the surveyor does, of course, work closely with an accountant, who ensures the necessary compliance with company law and accounting standards/taxation rules.

As another general business point, motivation, adrenalin and tempo is needed within a business looking to develop itself and this can be partly inspired by the consultant surveyor who has taken a directorship alongside other work. Key business skills, cultures and concepts include the appropriate division of responsibilities based on individuals' attributes, teamworking, trust, supportive people management, negotiating skills, financial awareness, the ability to read markets, and an ability to judge threats to the business (economic, legislative, competitive etc).

The investment company also had other key advisers in place, including a semi-retired, part-time consultant with accounting and financial expertise, and a small firm of lawyers. Importantly, the majority shareholders had the right vision and commitment to the business, including the willingness to appoint advisers of the right calibre — as opposed to less able, but considerably cheaper, personnel. Furthermore, "ensuring that people are motivated and well-rewarded" is often as important in dealings with professional advisers, as it is in attracting, motivating and maintaining staff.

As mentioned above, the right profile has to be created for the company, which benefits dealings with lenders, investors/shareholders etc, and also staff and professional advisers of the right quality. Examples of how this is achieved include the quality of company accounts and other information, professional offices, the use of international valuers, involvement of advisers with strong track records, a chartered surveyor as director, the growth story and the future aspirations of the business (including stock market flotation). To create a brand that attracts and retains tenants/ customers, key factors include targeted marketing, a good internal condition of properties, friendly management, flexible lease terms etc — in fact, all very simple concepts that help to secure voids of less than 5%. These were unrivalled by other investors who were less aware of brand value, and the need to "spend pennies to make pounds". As an example, £750 spent on improvements often secured an extra £5 in rent per week, equating to a 35% yield/income return

(£260 per annum ÷ £750), as well as generally providing an edge over competitors' properties.

A final business skills illustration is that property has developed into running a business. In contrast, much of surveyors' work is technically based, such as the judgment to apply a 6.5% or 7% yield in an investment valuation, and the related notion that property investment is broadly about 6–12% per annum rental returns, depending on the type of property, plus gains in capital value. Similarly, rent review, lease renewal, rating etc is relatively technical — although agency provides a good insight into the many operational issues facing businesses. With the thrust of university learning being mainly academic and conceptual, and many property decisions having to be justified by formal evidence, such as valuation, surveyors can suffer a mental straightjacket regarding the economics, dynamics, strategies and drivers within businesses, and how instinct and "just seeming to know" can often be the key to success.

Other companies, and other accounts

The above illustration comprises just one company and one set of accounts. However, as all companies are different, surveyors need to seek to understand the nature of the business, and the detail behind key elements within accounts in order to make accurate analysis and decisions (ie an understanding of business is needed, not just accounting numbers). The following section provides further information on accounts and their analysis.

Analysing company accounts

By providing only a grounding in *Business Skills for General Practice Surveyors*, and in view of the potential complexities associated with company accounting, undue detail is avoided (in some cases, this just means highlighting issues that may warrant further investigation in practice). As mentioned at the start of this chapter, the key requirement for most surveyors is to be able to broadly judge a company's current and future prospects. This involves identifying significant and sensitive items, and related commercial issues facing the business — rather than possessing a detailed working knowledge of accounting conventions.

Reasons for analysing company accounts

Accounts are analysed for numerous purposes, including by lenders, shareholders/ investors, purchasers of a business and also by the business itself as a management tool. Often, surveyors will analyse the financial standing or "covenant strength" of companies to establish their suitability as new tenants as part of agency/lettings work. When a tenant is seeking to assign a lease, the landlord will assess the prospective tenant's financial standing and judge, in relation to lease terms, whether this is acceptable. Investment valuations also consider a tenant's financial standing in establishing a yield — ie the stronger the covenant, the lower the yield and the higher the capital value. Surveyors sometimes work to crude methods of analysis, such as whether a company's profits cover the rent by three times. However, this is virtually meaningless in many circumstances (such a tenant utilising several properties), and tends to be a substitute for a lack of accounting/business knowledge when used alone. Surveyors will also draw on facilities such as Dun & Bradstreet, which analyse company accounts.

Having provided examples of a small property consultancy and a property investment company above, this section concentrates on the form of accounts relating to a retailing or manufacturing company. Many of the same principles of analysing company accounts are, however, followed with other industries/sectors, although key analysis will vary among sectors, companies, their circumstances and the purpose of analysis. This section concentrates on the analysis of small and medium-sized firms for the purpose of profitability, solvency/cash flow and gearing/borrowing, as opposed to investors analysing larger companies for the purpose of share price potential (although issues affecting larger companies are still raised).

Business understanding and wider issues

As well as drawing on accounting and financial information, an accurate judgment can only be made by also understanding the nature of the business, industry issues, and factors particular to the business, such as the imminent loss of key contracts, the recent award of new contracts, the loss of key personnel, distribution problems and gains from new technologies. Market trends and pricing in respect of raw materials, commodities, oil etc may be of considerable influence to the company's profitability. As illustrated in chapter 5, the impact of exchange rate variations need to be considered for firms significantly exposed to overseas markets (noting also that established overseas markets may be experiencing economic difficulties, thus influencing a UK company's turnover/profits).

Where companies are reliant on finance, an increase in interest rates will dent profits (see gearing and its effects later, and also pp14 and 36 in respect of property investment illustrations). For any business, competition may be increasing, serving to depress pricing and turnover, and therefore squeeze profits (which will be particularly problematic with small profit margins). These and many other factors again reiterate the importance of considering accounting information and its analysis alongside other factors.

As a broad indication, factors that will point to a business having a satisfactory outlook include stable increases in turnover, a steady cost base, solid profitability, good profit margins and a strong cash position without significant reliance on borrowing. The fact that a business is not growing is not a problem if the purpose of assessment is solvency/security. Many small businesses develop to a natural level, serving largely established markets/customers, and secure appropriate financial rewards. Growth may not be economically viable, not least because it can dramatically change the nature of the venture, the personnel needed, the level of costs incurred, etc. In other cases, while growth may be viable, and actually potentially very profitable, proprietors' personal preferences are to maintain the status quo. A family business, for example, can be keen to maintain family control, rather than introduce outsiders. Disaster can strike when the running of the business is handed from those who have built the business onto family members who lack the ability to run the business despite their many years' experience in the business, and tuition from others. Indeed, the calibre of a company's personnel is another key factor in judging businesses, and especially small businesses. If a business is contracting, again this is not necessarily problematic in terms of solvency/security — although the extent of contraction, the reasons for it, wider effects and the effectiveness of managing the process should be considered.

Furthermore, as shown in the example of the investment company, losses and a squeezed cash position can belie the growth and further potential of a company. New, sufficiently knowledgeable investors would see the rewards available and be keen to commit funds, whereas new lenders would be nervous and may not make a commitment (another illustration of how the analysis of accounts and also the conclusions drawn depends on who is analysing the accounts and why).

Variations in accounting formats

Although company accounts contain similar elements, the terminology relating to some entries may vary, as may the presentational format. A surveying consultancy or property investment business, for example, is different from a retailer or manufacturer in that there is no production and distribution process.

Set out below is a profit and loss account appropriate for a consultancy firm (earning fees and incurring expenses), followed by an illustration of a manufacturer, which introduces "cost of sales" and "gross profit". Examples of balance sheets are then provided. (As mentioned above, accounts follow statutory formats, including prescribed headings, but the examples are simplified in some aspects in order to concentrate on the key issues.)

Table 7.15 Profit and loss account for a consultancy firm

	(£)
Turnover	950,000
Administrative expenses	700,000
Operating profit	250,000
Interest receivable and similar income	10,000
	260,000
Interest payable	(30,000)
Profit on ordinary activities before taxation	230,000
Extraordinary charges	(40,000)
Profit before tax	190,000
Taxation	(35,000)
Profit for the financial year	155,000
Dividends	(65,000)
Retained profit transferred to reserves	90,000

For a manufacturer, gross profit would initially be calculated, and then net profit. Gross profit shows the profit on the manufacturing process, and then net profit shows the overall profit bearing in mind that the company has administrative, sale etc expenses. (In the profit and loss account above, net profit is effectively operating profit, but various categories/qualifications of profit need to be shown.)

Table 7.16 Profit and loss account for a manufacturing company

	(£)	(£)	(£)
Sales/turnover			500,000
Opening stock — raw materials	100,000		
Purchases in year	50,000		
	150,000		
Closing stock raw materials	40,000		
Cost of materials used		110,000	
Direct labour		70,000	
Indirect labour	25,000		
Manufacturing overheads	30,000		
		55,000	
Cost of sales			235,000
Gross profit			265,000

In the case of a retailer buying goods for sale, the cost of sales/gross profit would be similarly determined with regard to opening stocks, purchases and closing stocks, and could include any direct costs incurred in presenting the goods for sale. Rent, staff salary, overheads and advertising costs would be deducted from the gross profit to produce a net profit.

The balance sheet example below is designed to show the structure rather than detailed entries (such as a breakdown of current assets).

Table 7.17 Balance sheet — standard format

	(£)	(£)
Fixed assets		
Intangible: goodwill		20,000
Tangible: property		300,000
other fixed assets		30,000
		350,000
Current assets	50,000	
Creditors		
Amounts falling due within one year	40,000	
Net current assets		10,000

	(£)	(£)
Total assets less current liabilities		360,000
Creditors		
Amounts falling due after more than one year		60,000
TOTAL NET ASSETS		**300,000**
Capital and reserves		
Share capital	100,000	
Share premium account	25,000	
Revaluation reserve	55,000	
Profit and loss account	120,000	
SHAREHOLDERS' FUNDS		**300,000**

An alternative presentation, which is allowable but less commonly used, is shown below.

Table 7.18 Balance sheet — alternative format

	(£)	(£)
Fixed assets	350,000	
Current assets	50,000	
	400,000	
Current liabilities	40,000	
NET ASSETS		**360,000**
Financed by		
Share capital	100,000	
Share premium account	25,000	
Revaluation reserve	55,000	
Profit and loss account	120,000	
Long-term debt	60,000	
CAPITAL EMPLOYED		**360,000**

"Long-term debt" is excluded from the liabilities, which then form net assets, and instead is added to shareholders' funds in order to produce a total of "capital employed". "Capital employed" equals the "net assets" figure. The significance of this variation is explained later.

Assessing profit and profit margins

Chapter 5 explained the relationships between fixed costs, variable costs and profit margins, and outlined the sensitivity of changes in turnover and cost to profitability, together with economies and diseconomies of scale. Reference should be made back to the section "Costs" on p74. The points made were more in the context of a business evaluating its own position in order to make operational decisions — but others will be keen to determine that the business is trading cautiously given any prevailing sensitivities and risks.

The "gross profit margin" is calculated as gross profit ÷ turnover, such as £150,000 profit ÷ £1m turnover = 15%. Net profit margin is calculated similarly, and it is generally net profit before interest and taxation that is considered. However, where interest is significant, variations in the calculations could be undertaken in order to help show the sensitivity of interest payments to profitability.

As a general point with profits, accounts will separate out various elements that contribute to the company's profits. Such entries include "operating profit", "other operating income", "other operating costs", "extraordinary charges" and "items not reflecting trading operations". If the entries in the accounts do not provide adequate explanation, the notes to the accounts should do. The different entries enable the company to separate out the profit made from the actual trading activities/operations of the business from aspects such as interest received on bank account balances, interest payable on loans, investment income/rental received and income from shares/investments in other companies. EBITDA (earnings before interest, taxation, depreciation and amortisation) is also referred to and is a helpful measure of actual trading performance/profit, unfettered by factors that influence stated profit in practice.

"Extraordinary charges" or "items not reflecting trading operations", known also as "exceptional items", are items that are sufficiently important to be highlighted separately — such as the cost of redundancies, or defending a takeover bid (although research and development expenditure would normally be "capitalised" and entered as an asset on the balance sheet). By stating these separately, a more accurate representation of profitability can be shown in respect of the company's normal trading activities. In the example on p128, "extraordinary charges" was £40,000.

The phrase "above the line" or "below the line" refers to whether items have been included above the line in operating profit, or below the line as exceptional items. At times, companies may seek to include items below the line that ought to be above the line in order to suggest that the profitability of normal trading operations is greater than is really the case. If such items appear significant, the surveyor should investigate further.

The distinction between net profit/net profit margin with exceptionals, and without exceptionals, shows the potential profitability in the next and future years. As another example of considering the facts behind the figures, exceptionals may, for example, have been related to the closure of some of a retailer's poorly performing outlets. This could mean less contribution to turnover/profits, although in the case of loss-making stores, this would increase profits. Consideration also needs to be given to the difference between continuing and discontinued operations. A business may have ventured into non-core areas, incurred losses, and withdrawn from such markets. Sales of parts of a business or fixed assets, such as property, may be included in reported profitability, but turnover/profits relating to such aspects will not be included in subsequent years' accounts.

The trend of margins also needs to be taken into account. Falling margins could show a company to be losing its competitive edge, such as in the case of a market leader, or a niche concept that has lost its novelty value. Falling margins may, however, simply be a consequence of strong expansion. In the case of a small property consultancy, for example, growth from the format of one or two principals to a business employing a staff of 10 changes its economics considerably and, although margins are falling, absolute profitability is increasing. Nonetheless, the size of margins will still be important, with larger margins meaning that profitability is more defensive/less sensitive against changes in turnover and expenses/costs. Higher margins can attract competition, although barriers to entry to an industry provide protection — such as high capital costs or special expertise (although a company may simply enjoy inherent cost and trading/pricing advantages that account for its high margins).

It may be necessary to consider the impact of cost-cutting programmes: in simple terms, have the necessary efficiencies been secured for the better, or have costs been cut that are essential to maintaining current turnover and/or further growth? The redundancy of too many baggage handlers and other airport staff, for example, could cause chaos, which soon hits revenues/turnover, as well as reputation and brand.

Other measures of profit are shown later in the section "ROCE".

Factors affecting reported profit

By examining three or more previous years' accounts, the trend of turnover, costs and profit margins can be analysed. Expansion, as inferred by increased turnover, could be due to genuine expansion reflecting the success of an increasingly profitable business, or it could be the result of price-cutting, which decreases profits. A retailer, for example, may have misjudged seasonal sales, ordered too much stock and had to discount heavily in order to cut losses. One mistake in one year does not necessarily mean that there are inherent problems with the business. However, retailing and leisure markets can experience fickle consumer behaviour. A failed sales campaign or badly judged adverting theme, could lead to costly problems, as well as representing a misunderstanding of what consumers want. Again though, financial standing/covenant strength may still be substantial, with a retailer being able to ride out temporary difficulties even over a period of years. Even if profit performance is poor, the net assets position could be strong, and the company may be able to secure a successful sale or merger if necessary. When assessing company accounts, the effects of inflation need to be considered. Approximately 3% growth in turnover, for example, may simply mean inflationary price increases and no rise in sales volume (or turnover in "real" terms — ie net of the effect of inflation). A 5% reduction in cost, given the same sales volume, could mean a net achievement of 8%. (Inflation generally has a progressive effect over the year, and as the previous years' figures were derived similarly, a comparison at the approximate annual rate of inflation is valid.)

When assessing companies' accounts and general fortunes, surveyors should be alert to any matters untoward. For example, companies may seek to prop up profits for a relatively unsuccessful year, and generally wish to show progressive growth, as opposed to gyrations in performance and profitability. Turnover includes debtors, comprising work that has been completed and invoiced (but payment not received by the year-end) and work that has been completed but not invoiced. Where a company's turnover is relatively frequent, such as numerous daily cash sales with a shop, there is relatively little scope for turnover to be manipulated in terms of the year it falls in company accounts. However, where invoicing is relatively infrequent, such as for work undertaken over a longer period of time, or large single items being sold, there is more scope for a company to allocate turnover to a particular year. A struggling company may, for example, include as much turnover as possible, and as few expenses as possible, in order to make profit artificially high. If there were problems relating only to the trading year, this would help even out the trend of profitability but, for a company in continuing decline, this would make the next year's figures even worse. The company may also trade in a certain way around the year-end, with an eye on the most favourable accounting position that can be created. Property sales prior to the year would usually generate income and improve the company's cash position, and perhaps also profitability (although if values had fallen and the property had been mortgaged, the opposite may be the case).

With reference to the illustration in respect of cost of sales and gross profit on p129, profitability can be influenced by the closing stock figure. In the example, there is little scope for the company to alter profit in such a way as stocks are carried at low levels, but the effect could be particularly significant where a retailer works to relatively high stock levels in order to optimise the range of goods for sale. The time taken to sell goods and the mark up on cost/profit margin would also be relevant.

Although work in progress is not included in turnover, such as surveyors' work undertaken to date in respect of lettings and sales that will not secure a fee until tenants/purchasers are secured, costs will still have been incurred in connection with such work (ie salaries and other costs). Such costs can be excluded from the usual costs in the profit and loss account, thus increasing profit — with the same amount being treated as current assets in the balance sheet. Similarly, if such costs are not accounted for in this way, profit can be recorded at lower amounts. As well as any benefit in smoothing out profits, lower profits will reduce tax liability for the year.

Cash businesses can also have lower profitability in company accounts than is actually the case, because the owners do not enter all sales/turnover in the books in order to reduce profit and therefore tax. This hinders any sale price achievable for the business, as well as making finance more difficult to raise, and possibly more expensive.

Individual items of expense can be examined to assess their impact on the fortunes of a company (such as if wage/salary rates were increasing in a particular sector). Expenses can be categorised in company accounts, such as "staff costs", "depreciation and amortisation" and "other operating charges" in the investment company's profit and loss account on p113. A manufacturer might categorise expenses as "selling", "distribution", "financial" or "other". The section "Accounts examination — profitability and remuneration" on p122 illustrated the need to scrutinise profitability in relation to the way in which shareholders'/principals' remuneration is provided. As also mentioned above, it is likely that more detailed management accounts are needed, with statutory accounts not necessarily providing the necessary level of detail, as well as being only a snapshot in time.

A way to ensure that accounts can present a good impression would be, for example, a property investment company having a year-end of, say, 31 May, knowing that rents would be collected on a quarterly basis, and the cash position should look relatively healthy as there is time to secure the rents for the 25 March quarter. In contrast, a 31 March year-end may mean that debtors are high.

If necessary, companies can extend their year-end. If companies are likely to have poor accounts, they may seek to delay filing at Companies House, and therefore public availability as long as possible, even if this means incurring a fine. This could provide the extra time for the current year's trading to be shown within the financial statements, limiting the effect of the previous year's results. Such matters are usually in connection with stock market-quoted companies, and the actual delay in filing accounts could, in itself, cause a share price to suffer.

Organic and acquisitive growth

Organic growth is where a company grows naturally through its success, such as a retailer opening outlets in more towns. Acquisitive growth is where the company instead acquires other companies. The likely success of acquisitions and mergers therefore may need to be evaluated. Sometimes two businesses will enjoy great synergy (ie one plus one equals three or more), whereas others will not be an appropriate fit. Even where there is synergy, acquisition and merger may be poorly managed, with clashes of cultures and operational difficulties preventing the full potential to be exploited. Redundancy programmes and property relocations/ mergers are just two examples of many possible problems. Acquisition and merger decisions made by a company's senior personnel can sometimes reflect the greater empire, power and pay thought achievable, as well as the enhanced exhilaration compared with day-to-day work. Also, an eye can be taken off the core running of a business, causing profitability to suffer. Another aspect of growth is its pace. A business can grow too quickly and lose control of its various processes.

Goodwill

If a company buys another company, "goodwill" is the amount paid above the net asset value (assets less liabilities) of the company. As a basic example, a surveying business can be highly profitable without fixed assets, and although there may be a small net asset value, its sale price would reflect the level of profitability achievable by the new owners. When one company buys another, goodwill needs to be entered on the balance sheet (ie capitalised) and is then written off to the profit and loss account (and decreases as an asset on the balance sheet) over a period of time that reflects the useful life of the goodwill to the company acquiring the goodwill.

Companies cannot capitalise goodwill on their balance sheet because the potential sale value of the business is worth more than its net assets. This is another example of why a company's balance sheet does not necessarily represent the actual value of a company or its individual assets.

Depreciation and related issues

Fixed assets are depreciated over their useful economic lifetime by making an annual charge to the profit and loss account (and a corresponding deduction to fixed assets in the balance sheet). All tangible fixed assets should be depreciated, as per the Companies Act. Motor vehicles, plant and machinery and other such assets genuinely depreciate in value, and an entry in the balance sheet after depreciation is a reasonably accurate reflection of a company's assets. In contrast, with property, although depreciation/obsolescence is experienced, values will usually still rise over time, and property does not necessarily have to be depreciated (see p141).

Depreciation can be provided by the straight line method, eg £1m cost or over 10 years = £100,000 per year) or the reducing balance method (year 1: £1m less 20% year 1 = £800,000, year 2 £800,000 less 20% and so on). Resale value could be incorporated in the calculations, such as £1m less £300,000 resale value after 10 years = £700,000, which is then depreciated as above.

By entering depreciation in the profit and loss account, this reduces profits. It does not, however, necessarily reduce taxation, as there are specified rates known as "capital allowances", which can be set against profits/taxation. Where capital allowances exceed depreciation (such as a first year's allowance of 40% compared with a depreciation charge of, say, 20%) a charge to deferred tax would be made, representing the tax on the difference. This would be credited to the "Deferred taxation reserve" that appears on the balance sheet under Provisions for charges and liabilities. The taxation charge in the profit and loss account would therefore consist of the actual tax payable for the year plus the charge to deferred tax. Conversely, where depreciation exceeds capital allowances (such as where ongoing capital allowances are less than depreciation owing to the impact of high first year allowances in a previous year), the Deferred taxation reserve would be debited and the profit and loss account credited, thus reducing the overall charge against profits for the year. Deferred taxation could relate also to other matters where tax liability is provided for in accounts, but is not currently due (see "Property", p141). The actual tax due is, of course, shown in the profit and loss account. Also, the amount of taxation in company accounts is not necessarily a straight deduction at the appropriate rate for corporation tax, as some items are treated differently, such as entertainment and hospitality not being tax deductible, or the cost of remediating a contaminated site benefiting from a 150% charge against turnover/profits (as opposed to the usual 100%). Taxation may also be influenced by tax losses available (from losses made in previous years).

With companies whose business involves a relatively high requirement in respect of capital expenditure/investment, the treatment of depreciation could have a significant affect on reported profitability. Consideration would also have to be given to the lifespan of assets, and the need to invest in new assets (ie assets nearing the end of their life would see low, if any, depreciation in the profit and loss account, making profits relatively higher, but capital expenditure is needed in order to maintain profitability). If depreciation has a significant effect on a company's accounts, an accountant's advice should be taken.

Liquidity

"Liquidity" or "solvency" ratios assess the ability for a company to remain solvent. Although this primarily means its ability to not go bust, it also means its ability to avoid any further financing.

As previously mentioned, "fixed assets" comprise "tangible" fixed assets, such as property, plant and machinery, vehicles, fixtures and fittings; and "intangible" fixed assets such as goodwill and patents. Current assets include debtors, pre-payments, cash/bank account deposits and stocks. Stocks include raw materials, work in progress and finished goods, and are included in the accounts at the lower of their cost, or net realisable value. Current liabilities include bank overdrafts, creditors, accruals, hire purchase and other finance. Creditors, as part of current assets, generally include amounts owing in less than one year. Creditors also include corporation tax due and dividends that have been declared but not paid.

The "current ratio" is current assets ÷ current liabilities, and is also known as the "working capital ratio". (The actual working capital is current assets less current liabilities, known also as net current assets, or net current liabilities as the case may be). A current ratio of 1 (100%) shows that liabilities could be met. However, the working capital, or "cash flow" position may still be poor. Stocks, for example, take time to be turned into cash, and are not sensibly sold as raw materials or work in progress. Money from debtors can take time for the company to secure. Although payments could be delayed to creditors, creditors may pressurise the company for payment, and even commence winding up proceedings. The respective breakdowns between stock, debtors and cash will help a more accurate view to be taken, but as a guide, a ratio of 2 (200%) shows reasonable financial strength, whereas 1.25 to 1 is cause for concern. Again, wider factors need to be considered, noting, for example, that longer-term borrowings may be unduly high, and help provide strong working capital — with longer-term borrowings being listed in the balance sheet as "Creditors — amounts falling due after more than one year" (ie not current assets). This is also a good example of why the various accounting ratios and other statistics need to be considered alongside one another. As another example, the working capital ratio may be low, but not problematic as there is a strong capacity to secure long-term borrowing if need be. Attention would also have to be given to the natural cash flow of the business, as this will vary between trades/sectors.

The "liquid ratio", known also as the "quick ratio" or the "acid test ratio", is current assets less stocks ÷ divided by current liabilities (reflecting the points made above regarding stocks taking time to be turned into cash). Different phraseology can appear in literature, and there are variations in the aspects included or excluded in the calculations. The comments in this chapter show why there is a phrase "cash is king!"

As an example of the need to scrutinise the detail, in the illustration of the investment company above, current assets of £148,121 comprised £119,232 debtors (ie rents due) and £28,889 cash at bank. "Creditors — amounts falling due within one year" (which is really "current liabilities" but without using such a heading) was £1,007,156 — suggesting a problematic position. However, the notes to the accounts show "Creditors — amounts falling due within one year" to comprise approximately £940,000 as mortgage loans and only £35,000 as trade creditors (together with approximately £30,000 representing rents in advance and national insurance, etc, owing). Other information showed the company to be trading, and growing well, with capacity to comfortably meet any liabilities from any pressing trade creditors. The amounts for loans were included in "Creditors/current liabilities" because such loans were repayable either within one year or on demand (ie although loans were technically repayable on demand, the likelihood of this being required was very remote). This shows why certain items may be excluded, and a series of ratios considered.

Borrowing and gearing

The possible effect of borrowing has been mentioned briefly, including with reference to interest payments and interest rates.

"Gearing" refers to the proportion of borrowed capital to total capital. This is really borrowed capital as long-term debt, but calculations of gearing can show all debt, which will be particularly important in smaller companies where the majority of debt is short term, and considered as current liabilities rather than long-term debt. As indicated, borrowing could, for example, appear as an overdraft in current liabilities (ie as a creditor), or may appear as longer-term debt elsewhere in the accounts.

In considering the affairs of the investment company above, total borrowing is approximately £2m. On the basis of the balance sheet on p114, net assets, which is the same as shareholders' funds, is approximately £1.5m. Total borrowing is therefore 133% of shareholders' funds. On the basis of the differently presented balance sheet, as on p130, if long-term debt, of approximately £1m, is included below net assets, and determines an entry of "capital employed", net assets (and likewise capital employed) would be £2.5m. £2m borrowings on £2.5 capital employed is gearing of 80%. The more surveying/property investment-related calculation of £2m borrowings against £3.5m fixed assets/property value shows the effective loan to value ratio for the portfolio to be 57%, which is low considering that the company can secure finance at loan to value ratios of 70–75%. This also shows that there is scope within the company for remortgaging against some of its properties, either to release equity to cover cash flow constraints (such as because of the refurbishment/repair programme) or to fund further acquisitions.

Comparisons between yields and finance costs would also be made, ie the margins are relatively high, both between the yield and interest/finance rate, and the inherent profitability of the business once the effect of the extensive acquisition and refurbishment programme works its way out. This demonstrates how high-yielding investments are relatively defensive against variations in interest rate charges.

The concept of gearing was also shown on pp14 and 36 when highlighting how property investors make their money. The principles are the same with other businesses and their financial requirements, in as much as higher borrowing/gearing increases the sensitivity of the relationship between variations in interest rates/finance costs and profitability. If things go well, gearing helps disproportionately higher profit to be made, but if things go badly, it can cause disproportionately higher losses to be suffered. Profit margins are also relevant, and the lower the profit margins, again the greater the upside potential, but also the greater the downside risk. High profit margins therefore provide a business with defensive advantages. In other words, although interest repayments may be a significant element of the total costs, they are not so significant in relation to turnover and profit. "Interest cover" shows how profit covers interest repayments, and is calculated by profit (excluding interest payments) ÷ interest payments.

As an alternative to borrowing, companies can draw on more capital from shareholders. This would avoid having to incur interest repayments, whereas dividends do not have to be paid on share capital. However, with smaller companies, shareholders may not have sufficient further capital to inject. Furthermore, if new shareholders are approached, and new shares are issued, they would dilute the returns available to the existing shareholders (assuming profits were made from the increased capital, as increased share capital and shareholders would similarly dilute losses). This is the same in large companies, with the company and its shareholders faring better by raising long-term debt than by the issue of new shares. The issue of new shares (or more precisely, the balance of ownership of shareholdings) also influences the main shareholders' level of control over the business.

In assessing a company's fortunes regarding borrowing, focus can be given as to whether there are assets that can be used as security for further borrowing and the overdraft position and its limit. Financial difficulties may only be temporary, whereas in other situations, will be indicative of the imminent collapse of the company.

A distinction is sometimes drawn between long-term debt/creditors (which incurs interest repayments) and short-term creditors, such as trade creditors and taxation due (which does not incur debt). However, in the case of smaller companies, loans and overdrafts, which are classed as short-term debt, would be subject to interest payments.

ROCE

As well as consideration of margins, profitability can also be measured as a "return on capital employed" within the business. This is only meaningful for businesses having a certain assets base that facilitates their operations, such as retailing or manufacturing, and is adopted more for larger companies. A professional services consultancy, such as surveying, may have a limited need for capital, and enjoys high profitability because of its people, as assets, rather than premises, plant and machinery etc.

ROCE is the profit ÷ the capital employed in the business, expressed as a percentage. The profit figure used should be profit before interest and taxation, but other profit figures can be used. The capital employed figure is really shareholders' funds (which includes share capital, reserves etc) plus long-term debt. Another measure is the profit ÷ shareholders funds' (and although termed return on shareholders' funds/ROSF may be referred to as ROCE). The accounts on p130 show how accounts can be presented differently — in one case showing net assets/shareholders' funds, and another treating long-term debt differently. This is an example of how scrutiny of accounts and the readjustment/ reordering of entries (and recalculation of ratios) may be necessary.

As shown in the accounts of the investment company, shareholders' funds are the same as net assets, and can be viewed as what a company owns (although note factors such as goodwill not being included, and properties not being revalued). However, if long-term debt is included in capital and reserves, and is added to shareholders' funds in order to show capital employed, capital employed shows how the company is financed. To demonstrate this, if in the case of the investment company, "Creditors — amounts falling due after more than one year" (£1,133,018) was included as long-term debt, net assets would be £2,669,492. Shareholders' funds would be the same £1,536,474, and capital employed would be £2,669,492.

Activity ratios

"Activity ratios" are used mainly to assess how efficiently a company operates stocking processes and debt collection, and the impact of this on cash flow, including the ability to turn stocks and debts into cash if need be.

"Stock turnover days" is the number of days taken to achieve cash from stocks, calculated by stock ÷ cost of sales × 365. The stock figure can be the average of opening and closing stocks, or if known, the average of stock levels during the year.

Again, wider issues need to be considered, such as the level of business and relatively high pricing of goods achieved if attractive credit terms can be offered to customers, and the substantial discounts available if buying stocks in bulk. As indicated above, stocks may prove unsaleable or have to be discounted but, on the other hand, stock levels need to be high enough so as not to create pressures in respect of production. Stock, as with debtors, has to be financed, either through interest payments or the interest foregone on cash holdings/deposit accounts.

"Debtors' days" is the number of days taken to secure cash from the point of sale (ie the point at which turnover is recorded). The calculation is debtors (or average debtors) ÷ turnover × 365.

Surveyors would not usually be involved in such calculations and, as such, further detail is not provided (although the principles show how surveying businesses can similarly use such measures in order to help optimise their own performance).

Group structures

The examples above concern a single limited company. Groups of companies and parent companies/holding companies and their subsidiaries involve more complex group structures, and accountant's advice should be taken if such situations arise in practice. This includes establishing the strength of a subsidiary, and obtaining parent company guarantees if need be (or a lease/contract being taken directly by the parent company) as opposed to a subsidiary being sold as a means of relinquishing liabilities.

Directors' reports, company updates and business plans

In the accounts of larger companies, updates from senior personnel often provide helpful information on the company's trading activities, and the future outlook for profitability, together with any particular areas of opportunity or risk. Reports for smaller companies will be more limited, and it will be necessary to ask specific questions of senior personnel.

For start-up companies, accounts will not be available, and for companies that have traded only for a short period of time, limited financial information will be available. Judgment as to a company's trading potential is likely to then be based on their business plan, which will include a detailed commentary on the nature of the business, its market and the background and track record of its principals. Financial projections are likely to be provided, together with information such as the capital committed personally by principals, and loan arrangements entered into with banks. Generally, if lenders or major investors are seen to be backing a venture, others can take some confidence in this when reaching their own decisions.

The writing of business plans is not covered in *Business Skills for General Practice Surveyors* as these will tend to relate to surveyors setting up their own business, or to the development of existing business activities rather than client dealings. However, further guidance is provided in *Starting and Developing a Surveying Business*, with guidance on the raising of finance and cash flow planning/ budgeting also being given.

Surveyors' internal company accounting

As well as surveyors having to understand accounting and financial information in connection with property dealings, an understanding of accounting and taxation issues also relates to the running of their business. In the case of property consultancies, for example, internal accounting systems will be in place to deal with departmental accounting, fees, salaries, bonuses, central overheads etc. Comparisons can be made in departmental profitability, and statistics calculated on fees per fee earner/surveyor, fees per employee (ie incorporating administrative/support staff).

Accounting and relevant analysis in the public and corporate sectors will be less orientated to fee income, unless, for example, there are notional fee arrangements in place based on the work being undertaken for a client/operational division. Management information will, however, still seek to assess the productivity of employees. Again, this is outside the scope of *Business Skills for General Practice Surveyors*.

Surveyors as share traders

Surveyors interested in trading shares can draw on some of the information included in chapter 5, chapter 6 and this chapter. The analysis of company accounts for the purpose of share investment will, however, warrant additional considerations not least because the judgment relates to share price performance, rather than only issues of solvency, gearing etc. An ability to read markets, and understand psychological and technical impacts on share price performance, together with other fickle trends, will be important.

Other examples

Many different examples of accounts could be provided, but the above shows the broad contents of accounts and illustrations of their interpretation. Copies of stock market-quoted company accounts can be obtained free of charge from the *Financial Times'* Annual Reports Service at *www.ftannualreports.com*. Details of directors' and secretaries' duties, and other information, is available at *www.companieshouse.gov.uk*.

Property

The treatment of property in company accounts is potentially complex, and this section sets out the basic principles. Illustrations of some of the issues were provided in the investment company's accounts, but this section provides a self-contained overview.

It is helpful to first consider the general position ignoring the new International Financial Reporting Standards introduced in January 2005, and which apply to companies listed on a European stock exchange.

Historical cost or market value

Owner-occupiers can include property in their balance sheet as fixed assets at the "historical cost" of acquisition, or at market value by revaluation. The choice has to be made in respect of all of a particular asset such as property, or none.

As well as the actual purchase price, historical cost includes transaction costs, such as surveyors' and legal fees, and also taxes, such as stamp duty land tax. The cost entry in the accounts may therefore exceed the actual market value of the property. Also, a bullish purchase price that a valuer cannot match, a special bidder paying a high price for strategic benefits and/or a general fall in market values, would account for a market value being less than acquisition cost. In practice, the company would not necessarily revalue so soon. The cost of refurbishments/extensions would similarly be "capitalised" to the balance sheet, although running repairs/maintenance would be included as costs in the profit and loss account.

Market value really means Existing Use Value (EUV), or for specialist types of property which do not trade on the market, Depreciated Replacement Cost (DRC) — as per the RICS Red Book with which surveyors will be familiar. Here, investment properties and surplus properties are valued to Market Value (MV). Although EUV and DRC is confined to existing use and ignores redevelopment potential, scope to realise marriage value, deal with a special bidder etc, such factors, where significant, should be highlighted in a valuation report, and could include a valuation.

Effect of revaluation and sales

Carrying property at market value, rather than at a possibly very low historical cost, increases the fixed assets in the balance sheet, and also net asset value and shareholders' funds. As shown in the example of the investment company, when the property is revalued, fixed assets and the reserves (the revaluation reserve) are increased by the same amount (or decreased if valuations are lower). This increased balance sheet strength makes finance easier to raise, and the company's financial position to appear generally more attractive. This is provided, of course, that property assets are of significant value in relation to total assets. Increases in the value of fixed assets/property also means that gearing is lower (ie the extent of borrowing compared with net assets/shareholders' funds/equity — see p137).

When the property is sold, the profit or loss on the sale is reflected in the profit and loss account, with profits being subject to corporation tax in the usual way (or losses being treated in the usual way). Taxation also needs to be considered when analysing property in company accounts, as capital gains tax may be incurred if assets are sold.

Profitability measures

In carrying property at a low historical cost, profitability in relation to a company's assets base looks attractive — as measured by ROCE — see p138. A consequence of including assets in company accounts at full values is that a relatively low ROCE can mean a company is considered to be making inefficient use of its assets, and therefore be regarded as underperforming. This can prompt other companies to consider making a takeover bid.

Takeover bids

A takeover bid can also be prompted by low asset values in company accounts, including as part of a prospective "asset-stripping" venture. Here, the true value of property becomes important in establishing a company's price for sale, merger or takeover — and likewise in defending a takeover bid. As mentioned on p27, this was shown in the takeover of Safeway by Morrisons and in Philip Green's bid for Marks & Spencer, where valuers were quickly appointed. Aside from such events, as stock market-quoted companies are valued primarily in relation to earnings/profits, a revaluation of property assets does not necessarily influence share prices significantly. However, property investment companies are an obvious exception.

Relevance of the balance sheet

Notwithstanding the above issues, carrying property in a company's balance sheet at market value, simply presents a more accurate view to account users. However, the many finer points of accounting, which mean that an accurate view cannot be seen without more detailed scrutiny of other information (including that which does not have to be made publicly available), include the capital gains tax liability realised on sale not being reflected in the accounts. Neither is account taken of sale proceeds being considerable less if disposal was enforced due to the company facing financial problems (nor of costs of disposal).

A balance sheet fails to represent a company's value for many other reasons, such as when earnings are strong in a business that is not reliant on assets. See also the section on goodwill, p134.

Investment properties

Investment properties are included in company accounts at market value (not historical cost). Investment properties are not just properties held by property investment companies, but also properties held by non-property companies as investments — meaning to hold to earn rentals or for capital appreciation. Properties that are temporarily not utilised by a company, and/or pending sale are not regarded as investment properties.

Leaseholds

Leasehold properties occupied by tenants at market rents (as opposed to long leaseholds at peppercorn rents and capital value/premium) have not traditionally had to be included in a company's balance sheet, with rentals being included in the profit and loss account along with other overheads/expenses. Reporting standards over recent years, including IFRS below, have increased the requirements placed on companies.

Sale and leaseback

As mentioned in chapter 4, p53, the financial and accounting benefits of sale and leaseback will also relate to the way that the company has carried the property in its company accounts.

This includes whether the property is included at historical cost or market value, how the sale/profits on property are accounted for, and how property influences balance sheet strength/gearing. Another point with reference to the illustrative market values on p54 of £1.5m vacant possession value and £2m investment value, is the £0.5m additional gain in value/profit realised on sale, which would not be in the accounts even if the revaluation of the property is up to date.

The timing of the sale and therefore the recording of profit/loss within a particular financial year may be important to the company. Also, if capital gains tax/corporation tax were significant, a later financial year would delay the actual payment of the tax liability. Another consequence of sale and leaseback is that the company loses an individual asset against which finance could otherwise be raised.

IFRS

International Financial Reporting Standards (IFRS) govern the way that property is treated in company accounts. The standards apply to companies that are listed on an EU stock exchange, and aim to harmonise European accounting practices. This does not include the many smaller companies listed on the UK's Alternative Investment Market (AIM).

The standards are published by the International Accounting Standards Board (IASB) and, as well as IFRS, are known as International Accounting Standards (IAS). Key standards are IAS 16 — Property, plant and equipment, IAS 40 — Investment property, and IAS 17 — Leases. These accounting standards apply to owner-occupiers, landlords/investors and tenant occupiers, and cover long-term property assets.

As well as affecting the work in which surveyors are involved as valuers, the standards, as with other accounting, financial and taxation matters, influence the strategic opportunities that property presents to the corporate sector. Property investment is also affected by the accounting standards.

The detail of the standards is, however, beyond the scope of this publication but reference can be made to RICS publication, *Property Under IFRS* (see *www.rics.org.uk*). Key issues are:

- Owner-occupiers can carry their long-term property and other assets at figures based on "original cost" or at "fair values" (which is equivalent to historical cost and market value mentioned above). Disclosure of fair values for owner-occupied properties carried at cost is only "encouraged". It could nevertheless become mandatory in the future.
- The choice of original cost or fair value also applies to investment properties. Disclosure of fair values for investment properties carried at cost is mandatory.

- The surplus or deficit on revaluation is charged to the profit and loss account under certain circumstances, thus influencing earnings/profitability (although not creating an actual tax liability). Variations in property values can make profitability more volatile.
- There are various rules in respect of leasehold properties, including the distinction between finance leases and operating leases, and leases being included on the balance sheet.
- Owner-occupied properties need to be depreciated irrespective of whether they are carried at cost or valuation. Investment properties need to be depreciated if carried at cost, but are not depreciated if carried at valuation.
- Financial ratios, such as ROCE and gearing, are affected by changes in asset and liability figures.
- There are increased disclosure requirements.

As with all accounting information, although certain ratios may commonly be considered, appropriate analysis of the component elements to accounts enables performance and risk to be assessed. Likewise, a true understanding of the business, its sector/industry and its prospects is vital.

With regard to surveyors' analysis of accounts, if the way in which property has been treated in company accounts has a potentially significant effect, an accountant's advice should be taken.

Negotiation Skills

Many areas of general practice surveying involve negotiation, and it is the good negotiators who are able to achieve the best results on behalf of their clients/employers. Property negotiations is, however, as much about understanding property issues and deploying commercial judgment, as it is about actual negotiating technique.

As well as property dealings, surveyors may also have to negotiate contracts with suppliers, financial arrangements with banks, and terms of employment for themselves or with their employees. Surveyors' professional relationships with colleagues, superiors, clients and others, while not necessarily being regarded as negotiation, will similarly be reliant on interpersonal skills and there will be points of strategic and tactical application.

This chapter combines comment on negotiating techniques with some of the issues arising within property dealings in which surveyors are likely to be involved. As well as providing a grounding in negotiating and related business skills, it highlights aspects that surveyors can work on in practice. Although this publication and training courses help develop surveyors' capabilities, practical experience is essential, including involvement in pressurised, and even hostile, negotiations. For example, one minor twitch of the mouth in the shape of an early smile, because the other side is offering something that is surprisingly welcomed, can alert the other side to its possibly misjudged offer, and result in something more favourable being sought.

Win-win scenarios

It is often said that negotiations need to be a win-win situation. Whether this needs to be the case depends on the nature of the negotiations, with some being one-off events and others having wider consequences.

In a rent review, for example, the parties are contractually committed, and there is scope for uncompromising positions to be adopted. Sometimes this will be with little regard to the preservation of the landlord and tenant relationship, with the fight for the best deal, and the financial gains it brings, being of overriding importance — such as to a private investor or small investment company in the case of the landlord. Although long-term investment performance is not generally well-served if the landlord and tenant relationship becomes strained, this is unlikely to be relevant for an investor looking to sell the property, and achieve the best rental settlement in order to help maximise the sale price.

As a general business point, surveyors, as with other professionals, who are willing and able to take a hard-nosed approach to business dealings can be highly valued by clients. In contrast, advisers preferring to remain in a comfort zone, and whose professional advice (even subconsciously) steers towards avoiding conflict instead of a more difficult, but commercially superior, option may be unable to command the same level of business. There are also surveyors who are particularly adept at dealing with sensitive issues through diplomacy and strong temperament, and are valued highly, for example, by clients needing to get out of difficulties.

Reflecting client Issues and objectives

Surveyors' handling of negotiations often needs to reflect the nature of the clients'/employers' business, and their profile within the market. A local authority, for example, could be conscious of political ramifications, and their tenants' ability to complain to a higher authority and possibly the local press. The corporate sector could be aware of the detrimental impact on their core brand if questionable conduct is reported from property dealings. A property company could place great importance on being seen as a fair and flexible landlord, because such an image in the market could lead to small business tenants introducing other businesses to the landlord.

Such factors illustrate the surveyor's need to understand the nature of a client's business, its property objectives, and any factors that may influence the way negotiations should progress.

At the outset of negotiations, in addition to the above points about the client's/employer's business, it is important for surveyors to determine what they want from the negotiations (ie establish objectives), including any particularly important elements of a deal. It could be necessary, for example, that a tenant secures sufficiently wide user clause provisions in the lease of a retail unit in a shopping centre, when it is the landlord's common practice to severely restrict the type of goods that can be sold.

Time is also needed to refine a strategy, including any negotiating tactics that will be deployed.

Prior knowledge

For most property negotiations, it is important that the surveyor gains sufficient knowledge of the issues before opening negotiations. An obvious example is to have researched comparable evidence in sufficient depth before making a rental proposal to a tenant in a rent review or lease renewal. Occasionally, there are instances where a surveyor recommends to a client tenant that a landlord's rent proposal should be accepted, because it is relatively low, and does not reflect recent deals that the landlord seems not to have picked up.

Surveyors also require due knowledge of their cases before undertaking negotiations, otherwise they will be unable to make sufficient points of response to the other side's arguments. The other side could also ask questions that the surveyor struggles to answer, placing the surveyor on a less confident back-foot, losing credibility and perhaps even being forced to concede too much.

As another example of the importance of being aware of the detailed facts within a case, including detailed lease terms, a surveyor may consider it preferable to delay the progression of a rent review on behalf of a tenant who is struggling financially in the belief that the tenant's cash flow position will be eased if able to pay the increased rent as late as possible (albeit backdated to the effective date of the rent review). However, if improved comparable evidence emerges during the period of delay, even though the valuation date of the subject review does not change, the rental settlement may be higher than if the opportunity had been taken to settle it earlier. Also, the lease may contain a provision for interest on the late determination of a rent review, and apply, for example, a particular clearing bank's base rate, plus 4%, to the increased rent, backdated to its effective date. Higher than anticipated rent, together with interest penalties, would be an unwelcome surprise for a client/employer.

Negotiation methods

Opening proposals are usually put in writing and follow-up correspondence entered into. It is important that written communications are of a high standard — not just in terms of content, but also written style. Articulate, fluent, well-structured and well-presented written material conveys credibility, giving the writer/negotiator stature, and helping to add weight to the points made. In contrast, letters that are undated, unsigned or have parts corrected in ink rather than being retyped can alert the other side to the possibility that the sender/negotiator has deficiencies, and might be ripe to take on strongly in negotiations.

In business correspondence, the signature is usually the only personalised element aside from printed text, and the right impression needs to be made in the eyes of clients and others. A neatly presented, suitably sized signature conveys conscientious attention to detail. A stylish element to the signature suggests creativity.

Telephone negotiations can sometimes be hindered by the surveyor's office environment, as conversations can be overheard, and possibly judged, by colleagues. As such, it may be preferable to work in a meeting room, from home or make calls elsewhere on a mobile phone.

Negotiations by phone can more quickly address the issues than continued written correspondence. Some surveyors take a hard line in print, but are more conciliatory once speaking on the phone. This may be because a surveyor has a relatively timid personality, a preference to avoid conflict, or just realises that to reach a settlement, compromise is needed. Negotiators can be conscious of the comments and general negotiating approach committed to paper, with some correspondence perhaps being drawn on as part of third-party/dispute resolution processes.

Surveyors can also be more amenable in face-to-face negotiations than in writing or over the telephone. The point in time therefore needs to be judged at which a meeting should be arranged, thus avoiding lengthy delays (although the continuation of correspondence presents a delay tactic if required). Video-conferencing is a possibility, although does not create the same environment as being in the same room as the other side.

Informal opinions

Surveyors should not generally give informal indications of their initial thinking as to values or other matters until the due market research has been undertaken. However informal and however qualified opinions may be, they stick in the minds of people and can be used as a form of psychological benchmark thereafter. Nevertheless, it can sometimes work well for a surveyor to say to a tenant, for example, that because of the considerable improvement in market conditions and the rise in rental values since the last rent review, that a certain rent seems likely, with this being far in excess of market rents, and also in excess of any figure stated in a rent notice. The tenant is perhaps shocked and disappointed initially, but when a lower proposal is later received, this can be more palatable than had it been when the figure was initially mentioned.

This process of mooting something far more severe than is actually intended, in order to for the actual requirements to be better welcomed when proposed, is the same principle of deliberate government leaks to the media. If, for example, wishing to increase stamp duty from the current 4% for properties above £500,000, to 5%, the government might arrange leaks and general rumour-mongering to the press and media that stamp duty could be increased to 6% or higher. This can work because some people are fooled into thinking that things could have been worse and the proposal is therefore acceptable, and those who are aware of the strategy have less scope to be vociferous. However, a challenge can still be mounted, with arguments against the actual proposal being reinforced by questions as to the credibility emanating from deliberately arranging leaks.

Starting points

In the case of lettings and sales, asking rents and sale prices that are too high might, in theory, provide scope for negotiation. However, when prospective tenants and purchasers have various options to consider, they often quickly dismiss the overpriced possibilities, and do not wish to incur the time and cost of establishing the extent to which rentals and other terms are negotiable. Also, non-property people can be unaware of the extent to which quoting prices for property interests are negotiable, and some do not realise that there is scope for negotiation at all. This could see a tenant pursue relocation at lease expiry, for example, rather than negotiate on the rent.

Surveyors involved in agency/marketing work will need to establish the right pricing tactics to suit the circumstances. Guide prices at auction, for example, will often be some way below the reserve price (the minimum acceptable sale price) agreed between the auctioneer and the owner. In contrast, for house sales, the sale price will rarely be above the asking price.

In the case of rental negotiations for rent review and lease renewal, it is the comparable rental evidence that forms the thrust of negotiations. The landlord's initial rental proposal and the tenant's rental suggested in response may count for little at this stage. It is market practice for a landlord to quote a rent that is on the high side, and provide room for negotiation with the tenant — as well as perhaps being beneficial in terms of third-party determination if the parties cannot agree between themselves (see later). Also, for a property consultancy undertaking rent reviews and lease renewals on behalf of a tenant, it can, in fact, be beneficial if the landlord quotes a particularly high rent, as this provides more scope to acquire apparent savings, be more valued by the tenant client, and also secure a better fee (provided the right fee basis is agreed, ie a percentage of savings as an "incentivised" fee basis).

An unsupportable opening proposal by a landlord/owner might not be conducive to establishing a good rapport with the other side that will help negotiations proceed smoothly and amicably. Also, an unduly high opening proposal by a landlord could alarm a tenant into seeking professional advice, whereas with a more conciliatory proposal the tenant is more likely to negotiate personally, without professional representation. Similarly, the stance taken by tenants/purchasers is ideally supportable.

There are also situations where excessive initial proposals, such as through a rent notice, other legal notice or the commencement of legal proceedings, could see notices being held invalid, or the courts view a party's conduct unfavourably. This could affect an award of costs, and perhaps also influence the judgment detrimentally (as well as the potentially severe consequences if certain notices are invalid).

As well as establishing a starting point for negotiations, it is important to determine the extent to which negotiations can move from such a position. Clients/employers need to be appraised of both the starting position and the alternatives that might be acceptable. It is preferable not to promise a client/employer too much, only to disappoint the client when this is not achieved. This is another reason why thorough research and "knowledge before negotiation" is important.

The scope to deploy negotiating skills

It is sometimes said that negotiations should be about conceding a little at a time, working to the general spirit of compromise and, in the end, securing the required settlement. On occasions, it may be beneficial to state numerous requirements —

the majority of which are not really needed — in order to appear to be making considerable concessions in exchange for which the principal requirements can be secured. Some negotiations though, work on the relative basis of "take it or leave it", and the party with numerous points to debate will simply be rejected in favour of someone else able to reach an acceptable settlement more easily. Indeed, in strong market conditions, the landlord/owner could have numerous potential tenants/purchasers and simply invite best bids, rather than entering into negotiations on an individual basis.

Sometimes deals are lost because an owner/landlord and their surveyor judge that if the other side are raising certain issues at an early stage, then they are the type that will be forever raising issues, thus risking slow progress and the collapse of negotiations. It could be preferable to be open about key requirements from the outset, rather than become involved in an unnecessary game of "cat and mouse".

In other situations, the property dealings will be such that tortuous negotiations are the only way forward, such as the terms of a development agreement or a joint venture agreement between a developer/landowner and a regional development agency, or a range of terms in a conveyance needed to protect important adjoining operational interests. Another negotiating point is not to make concessions simply because the other side does, particularly as their starting point could have been tactically construed. The approach to negotiations will therefore vary considerably depending on the nature of the business being conducted.

Early stages of discussions

Most negotiations include a few minutes of general discussion, including introductions of the people involved or to kill time until others arrive. Gentle discussions about the state of the market, for example, could be relevant to the issues about to be negotiated, and surveyors will need to be on their guard. Similarly, surveyors should not feel the need to fill in any periods of silence during negotiations with further comment. As mentioned above, the surveyor should be aware of the case issues, and what negotiations should achieve, including any especially important elements. Points might have been listed in advance, although there might not be control over the order in which points are discussed (noting that the order in which points are covered/agreed could be of help or hindrance). It might need to be decided, on the day, who begins discussions, sets out their position, summarises the background or effectively chairs the discussion of points.

Finer points include having the required papers at the top of a briefcase, and not having to sift through A–Zs and other items. This would not matter if arriving early, but the possibility of late arrival needs to be considered, and others in the room halting discussions and observing the late arrival's movements in settling in. Also, a briefcase looks more professional than a travel-bag or rucksack.

In the early stages of negotiations, it is worth summarising points on which the parties are in agreement, and deal with any non-contentious points before moving on to points of greater significance.

Judging the other side's position

Clients/employers will not be impressed if deals have fallen through because an unnecessarily hard stance was adopted, or a negotiating strategy was simply misjudged. In a weak market, for example, a landlord/owner who suggests that there is a greater strength of interest and higher price aspiration than is really the case can see a prospective tenant/purchaser cease their interest. Even if the owner works hard to regain the interest, the other side can simply reject this. Surveyors have to judge their own position, and also that of the other side, carefully.

It is invaluable to determine the key requirements of the other side. Sometimes these can be relatively inconsequential to concede during negotiations, but a suitably presented "acceptance with great reluctance" can be beneficial for an overall settlement — especially when a grand concession makes the other side more willing to reciprocate when other points are raised. Another negotiating tactic is to agree to items provisionally, and to then to bring them back into discussions in order to help sway the other side in to meeting demands on other points.

It can be helpful to find something to which the other side are not entitled, but which they really need, such as a variation to lease terms at lease renewal for a tenant, but which the court would not grant. The other side's key requirements, and also general motives, might emerge only during negotiations. The initial exchange of pleasantries at a meeting, or any seemingly informal and/ or irrelevant parts of negotiations can produce information, which, when pieced together, can be revealing. As a further general negotiating point, surveyors need to guard against making general conversation that reveals information that is best withheld. For example, if the other side openly asks: "So what is your client looking to achieve from these negotiations?", the surveyor needs to be sufficiently versed in suitable responses. The quick-thinking surveyor, rather than being caught out, can use this as an opportunity to make comments that benefit his position.

Surveyors must also ensure that a favourable deal is not lost simply because they are too conscious of their personal standing in the eyes of their opponent or others through not taking a harder line.

Pointing to other options

In certain negotiations, it can be helpful to show that there are other options available. This will apply particularly to lettings and sales/purchases, where there are likely to be similar properties on the market. At lease renewal, details of other available properties should be sought by tenants — not only as comparable evidence, but also to show the level of alternative available space on the market, and the deals achievable were the tenant to relocate.

The importance of the availability of other options is also well illustrated in the case of a developer having to negotiate the purchase of a "ransom" strip from the owner of the key access to the site. There are cases, for example, where a developer's only apparent option is to acquire the ransom strip, and the owner consequently has high price aspirations. However, the developer secures an alternative route, even if involving purchasing a house in order to create land for access, greater cost to build a longer road and losing value through reconfigured site design. The presentation to the owner of the other option "changes everything".

Commercial imperatives and reputation

There are sometimes commercial imperatives to conclude transactions. A retailer might just have to accept onerous lease terms in a competitive market if they are to roll out their format through rapid expansion. A property investor will wish to be known as someone with a track record in providing a quick "yes" or "no" if properties are put to them, and in concluding deals speedily. In contrast, the investor with a reputation for seeking price reductions once due diligence on lease terms and title is undertaken is unlikely to be invited to bid when there are future opportunities (as well as losing the current deal).

Developers with a reputation of putting in high bids in order to be the successful bidder, but then seek price reductions in view of planning, contamination or other ground conditions will find themselves no longer on agents' mailing lists, and owners rejecting any bids made on other properties. The surveyor also needs to be streetwise to such tactics deployed by developers or others — such as being able to return immediately to another bidder, rather than appear to be left, in isolation, to deal with the developer seeking reductions. As already mentioned, it is helpful if it can be shown that options are available — such as reminding the keen-to-negotiate developer that "we are talking to other people". It is worth noting as part of professional ethics, that it is not permitted to advise people, for example, that a bid of £500,000 has been received when this is untrue, but if a bid has been genuinely received, this can be stated in order to help generate improved bids from others.

Handling more experienced negotiators

Sometimes a younger or less experienced surveyor will have to negotiate against a more senior surveyor. It is difficult when the other side considers that their superior experience and standing within the profession ought to see a deal reached quickly and successfully against a minion. Points might be put forward that lack merit in the speculative hope that the less experienced surveyor will be unable to mount a challenge.

It is important that younger surveyors do not try to create the impression that they are more accomplished and knowledgeable than they really are — not least because the other side is likely to see through this once discussing the detailed issues. Such pretence of greater standing could, however, work well if negotiating against non-surveyors — and some people are able to come across as particularly confident and knowledgeable in a situation, despite a relative lack of familiarity with the issues. Their key reliance, though, is the even greater lack of knowledge of others. Graduates, for example, can come across very well to non-property clients and take an active part in meetings, but their employer would hold them back from direct dealings with other surveyors and knowledgeable clients. If surveyors are out of their depth, this risks causing frustration with the other side, especially if it hinders progress. The other side might complain to the struggling surveyor's superiors, or deal directly with the surveyor's client.

Younger surveyors have greater scope to inform the other side that they will have to check out a certain point being made, or of the need to obtain their manager's approval to elements of a settlement. All surveyors are able to remain short of concluding agreement due to the need to make recommendations to a client, or discuss things through with a client — although, in practice, it will often be a formality that the client's acceptance is provided. In the case of in-house surveyors' roles, the need for superiors' consent, or the requirements of an operational department, can be helpful to negotiations.

Clients can, however, disagree with recommended courses of action, or wish to take a particular direction with property dealings. Difficult clients can make it hard for the surveyor to retain credibility with the other side. In fact, another way for surveyors to deflect difficult situations, or persist in obtaining particular requirements, is to express that they emanate from the client.

As another general negotiating point, surveyors can ask the other side about their scope to reach settlement without wider authority, and what any wider processes might broadly entail.

Conduct during negotiations

People with similar personalities, working in the same sectors of business and/or having common ground regarding external interests are more likely to enjoy each other's company, than the wrong type of contrasting personality. A young, brash, well-attired and overly well-spoken West End agent could struggle to make progress with the more down-to-earth personnel within northern local authority if the underlying attitude is one of superiority, and the expectation that there is a deal to be had — and one which will be in their favour. Surveyors dealing directly with businesses at the lower end of the market can be seen as officialdom, and an expensive car and sharp suit might not be appropriate. Negotiations with businesses and non-surveyors are often very different to negotiating with surveyors, with time being needed, for example, to explain technical aspects in layman's language.

For the individual surveyor handling negotiations, a belligerent approach might not find favour with the surveyors/firm against whom negotiations take place. Future transactions and negotiations can then be harder to conclude, and when generally phoning around the market for comparable evidence, a lack of co-operation could be experienced among other surveyors.

Surveyors may have to deal with awkward negotiators on the other side, including businesses and non-property people, as well as surveyors. They might make verbal agreement one day, and retract it the next. Representatives for the other side could have been on a negotiating skills course, and be trying to put a range of clever skills into practice. The overtly tactical, as with the belligerent negotiator, can cause more damage than gain.

During negotiations, it can be effective to make notes while an opponent is speaking. If, for example, a question is being asked, in jotting down a few words, a structured response can be prepared, and greater thinking time can also be facilitated. Also, if the other side hits on a point that has been unexpected, or challenges something to the surveyor's discomfort, it can be handled more easily if eye contact does not have to be maintained, and instead the surveyor is writing notes and concentrating on the paper.

Sometimes, however, negotiations have to remain face to face, and the loss of eye contact could signify retreat. Also, eye contact should not be lost with the other side immediately on hearing something that is uncomfortable, or being put under pressure (including through revelation of untruths). Eye contact will sometimes need to be lost, in order to think through issues accurately, and not be distracted by the eye contact or body language of the other side. Eye contact also needs to be lost in order to demonstrate the need to consider the points raised (as well as saying something on the lines of "I will need to think about this").

Negotiators should generally react dispassionately to points made by the other side. It would not be appropriate to react with gratitude, nor become more upbeat in terms of the mood adopted, if negotiations seem to be progressing well, as this could cause the other side to reappraise its position. However, it may be appropriate to express disappointment more passionately than is really the case. Surveyors also need to remain composed in terms of their tone of voice, impression of impatience, and possible escalating aggression and loss of temper. Tactically engineered aggression can, however, work well in some situations. A tone of indifference can also be helpful, especially when the other side is making threats regarding the consequence of agreement not being met.

Negotiations sometimes deteriorate to a war of attrition, with the individuals involved unable to step back from the issues and make the right decisions regarding compromise and settlement. In such a situation, it is helpful to discuss issues through with a colleague, or someone more senior who is able to provide a more detached view of the viability of continuing to fight for particular requirements. A break during negotiations also works well, including allowing people to calm down, discuss issues with a colleague or call a client.

Surveyors should occasionally smile during negotiations, and generally adopt a friendly demeanour (except where a particularly aggressive and unfriendly approach is thought necessary to achieve the desired result). Acknowledging good points made by the other side presents at atmosphere of conciliation, as opposed to a stubborn reluctance to concede.

Threats and honesty

If threats are to be made, they need to be reasonably actionable. Threats that are not actioned present an advantage to the other side, especially if threats are a frequent tendency. If deadlines are given, timescales need to be reasonable and, as with other elements of negotiations, a fallback position/option is likely to be needed.

A threat of third-party determination cannot sensibly be an opening gambit in negotiations, as this can have a detrimental effect on relationships and also prevent the prospect of an acceptable settlement being reached. Some surveyors, in response to a threat to proceed to third party, simply call the other side's bluff and leave them to incur the time and cost on the appointment of an arbitrator or court application only to resume negotiations thereafter.

Similarly, final offers need to be final and, although there could actually be some further scope to negotiate, the continued stating of a final position can damage credibility and prolong negotiations.

It is important that evidence is put to the other side in an honest way during negotiations. With a deal, as comparable evidence, involving a rent-free period, for example, a landlord might wish to present the headline rent payable at the expiry of the rent-free period, rather than discount the headline rent to reflect the rent-free period. Surveyors should always seek verification of evidence. Even when speaking to both parties involved in a transaction, there can be a different interpretation. Information that is presented incorrectly as part of negotiations can damage the offending party's credibility — although this might still not affect their stance as to price aspirations. There is, of course, a risk that the other side discovers discrepancies, but does not make an issue at this stage, and instead reserves challenge until a formal third-party process is under way — done in a way that can damage the credibility of the other side in the eyes of an arbitrator/independent expert/judge. Poor conduct from surveyors can also see complaints made to RICS, and possible disciplinary action taken against them.

For the experienced and well-trained negotiator it is possible to judge, to some extent, whether someone is not being truthful — although some of the signals can appear contradictory. People who lie are more likely to look away and make reduced eye contact, but if aware that this is a characteristic are likely to consciously hold eye contact for an artificially long period. Lies generally involve greater concentration to ensure accuracy and consistency with other aspects, whereas the truth can be told more instinctively and articulately. Therefore, lies are more likely to be told with hesitation; however, if straightforward they could to be told more quickly in order to get the lie over with. A change in the tone of voice, especially a slightly higher, more anxious pitch, could indicate untruths, with other factors including inconsistencies, changing the subject, general discomfort, and touching the face. If questions are asked on the subject matter to which lies relate, responses can be hesitant and also inconsistent: bright people can think quickly, but less sharp individuals struggle to speak at their usual pace.

As a general negotiating point, the other side can be asked to justify their position on certain matters. Similarly, if proposals are dismissed by the other side, they can be asked to suggest an alternative.

Dealing with over-cleverness

Some personalities consider that there is advantage to be gained by power-dressing, and some negotiators believe there to be merit in arriving too early or late. Within negotiations, someone might frequently interrupt as a deliberate means of disrupting the flow of the other side. The difficulty in making progress because of undue interruption might need to be appropriately pointed out to the other side, even with reference to the approximate number of times such interruptions have occurred.

People will often feel more comfortable holding meetings in their own offices (as well as saving time, and therefore perhaps being more profitable for the business). Sometimes tables and chairs might be strategically arranged by the other side. There may be three or more people acting against only one. Here, one person may ask the questions, another take notes and another says nothing and only looks continually at the surveyor on the other side. If surveyors are themselves negotiating as a group, it is important to clarify respective roles at the outset.

It is vital that too much cleverness does not come into negotiations. The tone of some training guidance can be that the other side has not thought of the various tips and techniques themselves, and are not alert to the tactical approach of the other side. Relationships usually need to be amicable to produce the desired result (or at least for wider benefits), and overt tactics run the risk of alienating the other side.

Surveyors will not always have advanced knowledge of the personalities with whom they may be negotiating, and their likely behaviour based on previous dealings, so enquiries can be made with colleagues. RICS members' directory provides the dates that surveyors qualified, giving an approximate indication of age, and the surveyor's firm's website may provide a brief profile of the opponent.

Caution with friendliness

Surveyors should not drop their guard because of a particularly friendly manner being adopted by the other side. This includes not being induced into providing certain information, not reaching agreement on points without due consideration, and not being lulled into a false sense of security regarding the other side's promises, or any points agreed on a provisional basis. Also, whether the other side is being aggressive or friendly does not necessarily enhance or reduce the deal that may be achievable. Some people just find it good for their egos to be regarded as uncompromising. In contrast, a tax inspector may be very polite and professional during investigations, and conduct discussions in a welcomed good spirit, but still turn round and say, "that's £10,000 extra tax you owe us please!".

Contractual commitment

Surveyors are well placed to know that a deal is not settled until there is contractual commitment. Even then, there are certain property transactions and individual obligations which can be undone, including through the courts if need be.

During negotiations, people can agree to things that they later reflect on, and then decide against. Residential tenants can agree to take a property in the day, do their calculations in the evening and sleep on it overnight, to then advise the agent the next morning that it is not, after all, affordable and that they need to look at cheaper premises elsewhere. For house sales, prospective buyers can pull out because they find other premises they prefer, or because their initial excitement wanes.

It is important that momentum is maintained in reaching contracts. If lease or sale contracts are slow to be provided, and there are delays in negotiations, deals can fall through. Solicitors are ideally instructed in advance, with a draft contract being available to applicants as part of marketing, or at least when a bid is accepted. Surveyors' personal availability to take calls, office systems, contacts in the market and a range of other business issues and personal skills have a wider influence on the successful resolution of areas of practice that are subject to negotiation.

It is important at the conclusion of negotiations that the points of continuing dispute, or agreement, are summarised, and negotiations are not reopened because of confusion over the precise issues agreed.

As a general negotiating point, it is important that the other side is still made to feel good about the deal it has reached, even though this may not be the case. People need to feel that they are respected professionally, and defeat can be difficult to take — as well as sometimes storing up retribution for another day. Any triumphant celebration of a deal should not be seen by the other side.

Property market psychologies

Surveyors working in the residential market will see a greater range of psychological oddities than with commercial property. When the market moves out of a decline, for example, owners can refuse to sell if the price is below what they paid, and prefer to wait for the market to rise before selling. This is despite looking to move to a larger, more expensive house, which will have increased disproportionately in value compared with the existing property. Loss is emotionally negative (and also people can simply be too greedy). If a commercial property investor receives a "cash offer" of £100,000 from an investor with a good track record in completing transactions, and £102,000 from an unknown investor who needs to raise finance and is slow in providing information required by the owner, the £100,000 cash offer is likely to be accepted immediately. In the residential market, owners will sometimes hold out for £100,500 from someone in a chain, instead of taking £100,000 from a first-time buyer. This is despite the break in the chain enabling them to command a reduced price for their next home considerably in excess of £500. Similarly, homeowners are often against bridging finance because they see it as an unwelcome cost, and do not have the vision and understanding of the potential overall benefits.

As mentioned in chapter 1, an important element in business and within negotiations is psychology. Only once direct experience has been gained within particular markets, and by being among certain groups, can some of the fickle factors be understood. In the consumer markets, people can often "chase cheap", whereas in business there is a better recognition as to the balance between the price and quality of product or service. Even in business, there can still be a sway towards the cheapest option, such as a public sector department being rightly conscious of best value obligations and accountability/transparency. In response, businesses need the lowest possible cost base in order to maintain price competitiveness and optimise profitability — even if this means reducing the standard of service and quality/salary of personnel.

Third-party/dispute resolution

For some areas of property, if surveyors cannot agree between themselves, a third party needs to be appointed as part of "dispute resolution".

For a rent review, this will be an arbitrator and an independent expert, and for lease renewal will be a court process. Professional Arbitration on Court Terms (PACT) is also available for lease renewal, but is rarely used. Dilapidations settlements will be handled by the court, as will tenants' claims that a landlord has refused consent unreasonably and/or delayed consent in response to an application to assign or sublet (as per the Landlord and Tenant Act 1988).

Other areas handled by the courts include the pursuit of rent arrears, forfeiture of a lease following a tenant's breach of covenant, lease interpretation, lease rectification, boundary disputes, the enforcement by a landlord of a "keep open" clause, tenants' compensation for disturbance and tenants' compensation for improvements. Surveyors involved in construction and the procurement of works will draw on the services of an adjudicator, although disputes may still reach arbitration and the courts. Other processes, such as tribunals, are in place for rating appeals, capital gains tax and inheritance tax liability, and compulsory purchase compensation (although through appeal may reach the courts). Surveyors may also be subject to negligence claims, although insurers and their legal advisers may be in the lead in negotiations.

Areas related more to the surveyors' business include disputes with the Inland Revenue, Customs and Excise, suppliers in respect of service standards and overcharging and former employees, such as in respect of unfair dismissal.

In these and other areas of property and general business dealings, surveyors can gain great advantage from having sufficient knowledge of the dispute resolution process, and be able to convey a confidence and willingness to the other side in being able to take the step of third-party resolution.

Understanding the legal processes

An understanding of the processes associated with particular forms of dispute resolution will be important in protecting clients' interests. Examples include the service of a calderbank letter/offer at rent review, and the need to observe the Civil Procedure Rules (CPR) with lease renewal and dilapidations claims. A calderbank letter, for instance, can be an effective pressure tactic, as it concentrates the other side's mind on the wisdom of maintaining a particular negotiating stance, and reminds them of the potential award by the arbitrator of all costs against an unsuccessful party. If, for example, a landlord's rent notice is £50,000, the tenant's counter notice is £40,000, and the arbitrator's award is £45,000, the arbitrator may apportion costs equally (ie each party pays the half of the arbitrator's costs and bears their own costs). Had the landlord served a calderbank offer at £44,000, and the arbitrator awarded £45,000, the arbitrator may award all costs against the tenant because he had the chance to accept £44,000, and the arbitration and its costs have only been necessary because the tenant failed to do so. With disputes heard by the courts, "part 36 offers" provide a similar procedure. The importance of observing legal processes is illustrated by the CPR, as a party could win their case, but still be liable for costs because of their conduct, including their failure to observe pre-action protocols, and refusal to co-operate regarding alternative dispute resolution, which is favoured by the court.

An understanding of third-party processes, as with law and case law, is also important in view of its influence on the subject matter being negotiated. If, in a rent review, a landlord of an industrial estate is presenting comparable evidence only of the best transactions, the other side needs to be aware of the rules on "disclosure", and how an arbitrator might force the landlord to provide details of all lettings. Rules on what an arbitrator can and cannot do, and how certain types of evidence should be treated, can provide effective negotiating points.

The surveyor who has the confidence to take a case to a third party tends to be able to conclude cases more quickly than the surveyor who is nervous about such a prospect. This enables case work to be concluded quickly, and for fee rates per hour to be relatively higher, thus also enabling a surveyor/firm to be more price-competitive, win more business and generally be more profitable.

Evaluating the chance of success

Owners and their representatives usually wish to avoid third-party processes owing to their time and cost, and also the uncertainty of the result. There can also be perceptions in the market, for example, that arbitrators tend to split the difference, and tribunals for rating appeals are more likely to be biased towards the Valuation Office. Courts can be more lenient to tenants than landlords. Larger companies may also have less chance of victory than individuals, despite the legal principles being the same.

Sometimes it is worth having a speculative attempt at securing a result that does not seem possible, or even goes against established case precedent, not least because there may be differences between the circumstances of previous cases, and because courts and judges are not infallible.

Creating the wrong reputation

Larger property owners, and individual surveyors, also need to ensure that they do not get a reputation of never going to third-party determination. If, for example, a local authority landlord is viewed this way, and the market rental value of a property is £20,000, a tenant can simply refuse to pay anymore than £18,000, to see the local authority eventually agreeing to this, even though the initial valuation and minimum acceptable figure may have been £20,000. Another local authority, however, may benefit from a surveyor with experience in third-party processes, and a tendency not to hesitate to take such as step if need be. The ability for surveyors to refer to other cases they are taking to third party, or to third-party work in which colleagues are involved, can demonstrate that it is a step that the organisation/surveyor is prepared to take, and has the capacity to do so.

Wider consequences of actions

The above example illustrates a further point of considering the consequences of decisions in one particular case on wider interests. As well as market knowledge as to an organisation being soft negotiators, a lower rental achieved on one property (or within the same building) may be comparable evidence for further deals in the same portfolio (or building), thus compounding losses. The same principle applies when organisations consider it preferable to pay out compensation claims, instead of incurring the cost of fighting them, even though the claimant's case is spurious. The result is that the organisation begins to face a higher number of claims — many of which are inspired by the expectation that they will be settled out of court/tribunal, and some of which are simply speculative, or even engineered. Also, an individual decision maker within an organisation could find it more comfortable, from a personal perspective, to avoid the tension, and possible focus on their previous conduct, by challenging a claim through court/tribunal. Adverse publicity can, however, be an overriding reason to pay off a problem and, indeed, the threat against any individual or organisation of adverse publicity can be helpful to any negotiations.

Some property owners are reluctant to take the third-party route because its cost will come from a particular budget, whereas the difference in rent will really be lost elsewhere. Sometimes, the cost of commencing a third-party process is worth incurring as a way of showing the other side their confidence in the likely outcome of the case.

Dispute resolution and delay tactics

The dispute resolution process can also be used as a tactical weapon to secure beneficial delay (although this is not regarded as the most professionally desirable way of conducting business). For example, a tenant within a development scheme may take a particularly obstinate stance at the end of his lease regarding the landlord's attempt to secure possession. Although the tenant does not dispute the ability for the landlord to establish a ground of possession under the Landlord and Tenant Act 1954 (ground f — redevelopment), time may be needed to find new premises — and also to increase the landlord's anxiety as to when possession can be secured, when construction can commence etc. This induces the landlord to simply throw money at the problem, thus securing compensation for the tenant substantially in excess of the statutory amount available. This illustrates another point regarding negotiations, as well as being a general business point: that worst-case scenario should always be evaluated.

Other factors affecting property settlements

Within property dealings, there are many areas, other than negotiation skills, that determine whether a surveyor achieves a successful settlement. In valuation disputes, for example, the tenacity of the surveyor in seeking out comparable evidence will be important. If a tenant's breach of covenant can be identified, it may help a successful rental settlement to be agreed, on the basis that a blind eye is turned to the indiscretion. In the majority of landlord and tenant dealings, the preservation of relationships will be important.

With arbitrations, a party can sometimes be disappointed with a result, and it may even depart from both parties' expectation of market rental value, but it reflects the quality of the case put forward by the surveyor, including the relevance of comparable evidence, the valuation analysis and quality of legal support. The written and presentational standards of submissions could have an impact, as could the profile, experience and expertise of the parties' representatives.

Links with other business skills

Negotiation technique is only part of surveyors' necessary communication skills. The key requirement of effective communication includes the development of client and other relationships through strong interpersonal skills. Listening skills, assertiveness and an ability to influence people are all important. Report writing is another key communication skill. All such areas provide scope for further learning, including through dedicated texts — some of which extend to guidance on manipulating people, and the finer points of body language. Surveyors, as with other business subjects, need to be conscious of certain education and training options involving undue academic dissection of negotiating skills, which become remote from the issues and experiences arising in practice.

As mentioned at the beginning of the chapter, property negotiation is as much about understanding property issues and deploying commercial judgment as it is about actual negotiating technique.

Presentation Skills

Speaking in front of others is something that even the most experienced surveyors can dread. It can be particularly nerve-wracking and there is pressure to perform. Consequences of a poor presentation include damage to the credibility of the speaker/firm, loss of business and personal embarrassment.

Because many people prefer to avoid speaking to large groups, surveyors who master this challenge develop a confidence that gives them an edge over others. This helps accelerate personal development and makes surveyors more marketable in terms of internal promotions and new job opportunities. Also, as surveyors seek more senior positions, presentations become an increasingly common element within day-to-day work.

Types of presentation

Set out below is an overview of different types of presentation to various groups.

Small groups

Surveyors may have to speak to groups of only five to 10 people as part of internal meetings, sometimes in a managerial capacity. The surveyor will be judged by colleagues on how the meeting/presentation is handled, as well as on the actual information/instructions being given to the group. A manager may have to take a particularly authoritative stance with employees, or a surveyor may be presenting to people in higher authority and have to account for performance problems.

Large groups

When presenting to larger groups, the audience needs to remain captivated, and must be impressed by an articulate and informative performance. The size of the audiences could range from 20 people to hundreds of delegates in a conference room. All are sitting in judgment, and even if the presenter/surveyor is accomplished in the subject area and in delivering presentations, there is still pressure to perform well. Also, technology may need to be drawn on to enhance the presentation, but should not risk the presentation being seen as "style above substance".

Meetings, interpersonal skills, personalities

Meetings with a number of people similarly draw on presentation skills, with surveyors having to address the room on particular points. Surveyors may be asked difficult questions, and have to respond under the gaze of an audience (similar to questions at the end of a presentation).

Business pitches

As part of a pitch to win business, a prospective client may require surveying firms to give a presentation, demonstrating their understanding of the issues involved, and selling the reasons why the client should appoint them. There may be only a small amount of time to make the right impact and win over a prospective client who has not met the firm's surveyors, and may know little about the firm. As written submissions may not have been requested by the client, the firm's presentation performance could determine whether or not the business is won.

As mentioned in previous chapters, an understanding of a client's business and related property objectives is a key business skill. The same principle applies to presentations in terms of understanding the nature of the audience, and therefore the content and style that a presentation should adopt. In a business pitch, for example, an active demonstration of how an instruction would be handled, its identified objectives, illustrative solutions and expected results would be better than simply delivering a standard talk on the firm and its general activities. (Business development techniques are beyond the scope of *Business Skills for General Practice Surveyors* but, as a further area of study, surveyors should consider effective communication, interpersonal skills and psychological factors.)

As well as formal business pitches, more informal meetings, which lead to either winning new business or the failure to pick up a new client, can hinge on interpersonal skills, and how the surveyor focuses on pertinent points in a short time period. A prospective client may see a number of property firms as being equally capable, and an appointment could simply depend on the client's preference as to individuals and their personalities.

Initial considerations

The type of presentation undertaken influences the manner in which it is best delivered, and the extent to which notes can be drawn on, or a display/ PowerPoint can be used. A prospective client or panel of client representatives need to be looked in the eye, so the presenter will need a mental checklist of the points to cover, with minimal reliance on notes, and perhaps no PowerPoint facility. In contrast, a large audience expects the presenter to be relying on notes.

The structure and content of a presentation is also influenced by its purpose, and by the nature of the audience. Also, there may be particular requirements of a client, conference organiser or other host, which need to be included in the presentation.

The nature and knowledge level of the audience also determine whether style or substance will predominate. In the case of a business pitch to a non-property client who does not understand the detailed content, visual impressions and pure presentational quality may carry more sway. A more knowledgeable client is better able to judge property content and is unlikely to be hoodwinked by "style above substance". It is not, of course, always possible to know exactly what the composite expertise of an audience will be in cases such as business pitches. Indeed, as with negotiation and general business dealings, it can be preferable for a client to appear relatively lacking in knowledge, as this gives more scope for questions to be asked, and judge whether people are being accurate, or truthful, in their response.

Keeping to time

Practising a presentation will help establish the time it takes to deliver, which is important where the presentation needs to be of a specified length. Even if there is flexibility on the day, an event with several speakers will need to run to time, and it will not be appropriate to eat into the time of other speakers. If an event overruns, a small number of delegates can begin to leave the room — not necessarily through disinterest or protest, but because of travel arrangements and other commitments based round the scheduled end of the meeting. This is distracting for the rest of audience as well as the presenter, and prevents an event being concluded in the best possible way.

Easing into the challenge

An effective way to ease into the challenge of speaking in front of others and delivering presentations is to find straightforward opportunities where the consequences of any difficulties are minimal. A surveyor could arrange a talk to colleagues on a current issue facing the firm, or on a particular property subject as part of a lifelong learning/CPD training session. Graduates could undertake presentations to schoolchildren, or surveyors could provide a talk to university students. A local club or society may be pleased to hear the views of a surveyor on the property market and the increase in house prices.

As the numbers in the group can rise progressively, the surveyor gradually masters the ability and confidence to speak to large groups. If efforts are not made to obtain such initial experience, when an important presentation has to be undertaken as part of the surveyor's work, the nerves are likely to be intense (see below).

Familiarity with the subject

The greater a surveyor's knowledge of a subject and the more frequent the issues appear in day-to-day work, the sharper the mind can be when undertaking a presentation. In the same way that for physical tasks, training and practice help get the muscles into the groove, for mental tasks, training and practice gets the mind working in the required direction and with the necessary level of sharpness. Sometimes a surveyor may be working fluently in a paper-based way, but it is difficult to talk through the issues owing to a lack of familiarity with the verbal format of delivery. This is also why practising a presentation is important.

Presenters must also be concise, as opposed to using too many words to get to a point. The greater the familiarity with the subject matter, the fewer words people need to articulate their points. One aspect is the reduced amount of thinking time needed, and therefore reduced verbal padding while thoughts have to be formulated.

Breaking the nerves barrier

The nerves barrier is broken when a presenter knows from experience that, once they are on their feet, they will be able to get through the presentation and not be short of words. Until this stage is reached, part of nervousness relates to the fear of the consequences of nervousness, such as freezing on the spot. Another important element in easing into the challenges is therefore to practise speaking on issues without using notes. In developing the confidence and ability to talk relatively spontaneously, the surveyor is able to deliver presentations with more eye contact with the room, and less pre-occupation with notes and/or PowerPoint. This will be impressive to the audience — and in the presenter knowing this, confidence is increased yet further.

An audience knows that presenters will be nervous, and is unconcerned if presenters appear slightly anxious in the early stages. The occasional clearing of the throat and mis-use/immediate correction of certain words is not a problem, but it is important that the presentation is delivered articulately and fluently. Other consequences of nervousness include raising hands to the face too frequently, poor posture/positioning, clicking pens or tapping pens on a table while not conscious of the noise and irritation to others, fidgeting with paper and other items, and generally being unable to remain still. If a lectern is gripped too tightly, vibration may be heard through the sound system. A presenter therefore needs to be conscious of his actions, and learn to remain still (see "Positioning of the presenter" below). It may be necessary to clasp hands together.

A nervous presenter may also feel unwell, but this is just the effect of nerves. It is always important to eat and sleep well if experiencing stressful situations. Although adrenalin produced by the body should act as a stimulant, some presenters find it helpful to drink coffee/caffeine before speaking. It can also be helpful to take deep breaths prior to the start of the presentation (as this slows adrenalin and the pulse rate because the brain feels that sufficient oxygen is being supplied to the body).

Most presenters will, in fact, feel nervous to some extent, but this is an essential harnessing of energy and adrenalin so as to optimise performance. Once surveyors have broken the nerves barrier, they usually do not feel any anxiety, or need to find adrenalin, until 10–15 minutes prior to the presentation beginning. In all business environments, nerves should be seen as a power to perform. Some presenters will actually see a lack of nerves impede their performance, such as an experienced university lecturer delivering a dreary afternoon presentation to over 100 students.

Presenters will tend to be less nervous when their knowledge is superior to that of the audience. Nervousness can therefore be eased by the right mindset — with the presenter empowering themselves to perform by thinking about their high calibre track record and achievements, and being (at least inwardly) conceited in their superiority over the audience.

Nerves are also more intense where there are greater consequences of good performance. At a surveyor's key interview for a job, or a graduate's interview for the Assessment of Professional Competence (APC), surveyors are not nervous simply because of having to meet and present to a small number of people. If a surveyor attends a job interview on a speculative, indifferent basis, nervousness may not even be experienced.

A final point about nerves is that presenters can often report after a presentation how nervous they felt, including feeling hot in the face, and wondering how the presentation came across, but the person observing the presentation felt that the presenter's performance was accomplished and relaxed. Finally, it is always beneficial to arrange for the attendance of a constructively critical colleague at presentations.

Beginning a presentation

The beginning of the presentation is particularly challenging for the speaker. It is important to make a good early impact on the audience, and also overcome initial nerves.

An audience will be most focused on a presenter when they begin speaking. After around 30 seconds, some of the audience will be making notes or lose eye contact for other reasons, such as to view a PowerPoint or other display. Some people still concentrate well when not looking at a presenter, so the presenter should not necessarily view this as disinterest.

The presenter needs to be particularly familiar with the opening minutes of the talk in order to gain a fluent and confident start. Therefore, simple points need to be covered, rather than complex issues. The audience will welcome a brief overview of the scope of the presentation, its purpose and objectives, which are easy points for the presenter to cover confidently. Nerves will, in due course, ease and a more relaxed feeling will enable the more challenging aspects to be handled well.

Another point when beginning a presentation or returning after a break is to not begin speaking until the audience has settled. Several reminders may be needed before all of the audience cease talking.

With an audience of up to 15/20 people, the eye contact between the presenter and members of the audience is personal. Presenters need to guard against eye time being allocated disproportionately to one or a few individuals, and ensure that eye contact is shared in an inclusive way. As the numbers increase, it is easier to look around the whole room without making eye contact with anyone specific. Indeed, some presenters are less nervous with larger audiences, because they feel under less personalised scrutiny. It is also less likely that the presenter will be distracted by someone who appears disinterested or is fidgeting, or when shaking their head in disagreement with a point being made.

Scope for humour

Some presenters consider the most appropriate opening to be a light-hearted story that has nothing to do with the subject of the presentation. For the majority of business presentations, this will not be appropriate. It can come across as an unnecessary irrelevance, and a presenter seeking to win over an audience with something other than their calibre and the subject matter, and even hiding from the issues in hand. Humour might have its place elsewhere in the presentation, but should be limited.

It tends to be only a particularly charismatic and polished speaker than can inject humour in a beneficial way. Attempts at being humorous can also backfire, such as a lack of laughter from the audience, and it may be difficult for the presenter to recover. An audience can also begin to find amusement in the presenter's failed attempt to be funny. Rather than stories and jokes that have a punch-line for a laugh, a presenter can instead show humour from structuring comments to meet laughter if the audience sees fit, but which go unnoticed if not picked up by the audience.

Case examples and analogies are always a good way of retaining an audience's interest, as are direct links with their areas of work or other interests.

Selling points for the presenter

The presenter must consider what needs to be achieved from a presentation, and assess how to come across as impressively as possible to the audience.

In the case of a business pitch, for example, the surveyor may wish to show a prospective client the very detailed knowledge of market conditions and local issues, while at the same time illustrating an alertness to wider issues, which will add value to the client's interests. The more detailed the issues covered by a presenter, the more impressed the audience is likely to be with the presenter. A surveyor, for example, who provides a talk on "landlord and tenant case law update" is likely to be better covering three or four key cases in depth, rather than skirting over the issues arising in 20 cases. An obsessive and indulgent level of detail, however, needs to be avoided, with the presenter also considering a suitable level of detail for a particular audience.

Where presentations are undertaken as part of business development initiatives, such as contributing to a local RICS CPD event, care has to be taken that the presentation does not become too overtly commercial — especially as the audience are there to be informed, and may also have paid to be there. The presenter tends to best sell themselves and their firm through the subject matter, and very brief company details at the end, than through an overt sell. The use of case examples is also a more subtle way of selling the presenter and his firm.

Applying facts to practice

In the above example of the case law update, rather than simply providing technical/factual information, an audience would welcome illustrations of how new case law impacts on their work in practice. Here, examples can bring the presentation to life. A presenter may even pose a question or issue within case for the audience to ponder, before explaining the result of the case. Even without seeking responses from the audience, the posing of questions as part of a presenter's overall style is a good way of captivating the attention of audience — ie they quickly have time to consider a point for themselves, rather than information simply being delivered at them.

Sometimes a presenter can facilitate audience participation by posing a question and seeking responses. This can work well with the correct subject matter and the right audience, but can be a time-consuming way of getting points across and consequently needs to be limited. Audience participation is, however, important in delivering good quality presentations.

A presenter also needs to broadly judge from an audience as to whether the right level of content is being provided, and therefore whether there is scope to make detailed comment that appears to be welcomed, or whether it is preferable to move on to lighter, more interesting matters. Similarly, it is worth observing whether an audience appears in agreement with the points made, shown by nodding, or whether there appears to be some disagreement, shown by people shaking their heads or moving around in their seat.

Presenters should also use phrases such as "as you will be aware" to convey obvious points because they are either essential, there is uncertainty as to the underlying level of audience knowledge, or there is mixed audience of varying knowledge. Phrases such as "I am not sure if you are aware" should be avoided, and care taken throughout a presentation not to be inadvertently patronising or insulting.

Style of delivery

In addition to the demonstration of expertise, the presenter needs to appear suitably passionate, and deliver the presentation with the appropriate level of enthusiasm. This can be achieved by varying the tone of voice, and generally being articulate through a natural delivery of the talk (as opposed to being monotone). The tone of voice needs to strike the right balance between showing passion and professionalism, and the presentation being a hyperactive rant.

The speed of delivery will also be important. An articulate presenter can speak relatively quickly and captivate an audience, and this will be better than the pondering intellect who struggles to deliver the points with fluency. A slower pace is, however, necessary when the audience need to grasp complex points, and when needing time to reflect on points. Pauses also need to be suitably timed (see later).

It is important that the presenter "is themselves", and does not seek to elevate their stature by a more elevated, "put on" voice, or pompous style of delivery. Apparently authoritative, but actually ill-informed use of buzzwords, phrases and other facts can backfire.

When presenters rely too much on notes, the presentation can lack punch. Particularly poor presenters sound as if they are reading out the presentation word for word, which is a reason why the presenter needs to develop the ability to talk relatively spontaneously from key points. Skilled presenters, however, such as a television presenter working from autocue or a politician with a full script, can be reading word for word without the presentation/speech sounding read out. If surveyors are utilising a script for the whole or part of their presentation, the text needs to be prepared for the spoken word, not the written word. For example, "furthermore" is effective in print but is not really used verbally.

In some cases, surveyors may need to work closely with a full script, such as when dealing with sensitive issues, or where there is scope for misinterpretation — including by the press who may be in attendance. A copy of the speech and an accompanying press release would be important in ensuring accurate reporting, and such material can also be available to the audience (as well as anyone not in attendance, for posting on a website, for example).

Structuring a presentation

The above points, as well as the presenter's confidence and ability in delivering presentations, will influence how the presentation is structured. It is important that the individual presenter adopts a format regarding notes, PowerPoint and presentation style with which they feel comfortable — and not necessarily be coerced into a particular format by a training adviser.

Beginning, middle and end

It is sometimes said that a presentation has a "beginning, middle and end" and that a presenter should "tell the audience what they will be told, tell them, and tell them what they have been told". Although simplistic, the opening remarks and the final comments are important, self-contained sections. Whether tips such as "make no more than five points" are appropriate, depend on the nature of the presentation, such as a 10-minute talk on a property development or an hour-long summary of recent case law.

Structuring into sections

A presentation needs to be broken up into sections. This will give the impression to the audience of the presentation being well structured, and will also make the presenter's task easier. The end of a section of gives the presenter time to pause for breath, and gather his thoughts if need be. The pauses between sections will be for a longer period than the brief pauses within sections, and this can enable an audience to refocus their attention. This can be reinforced by the presenter announcing the heading/nature of the new section, and the technique could also be used to stress the importance of a particular key section within the presentation. Headings within PowerPoint presentations and/or notes should correspond with the presenter's comments, such as under a section of "Economic conditions", the presenter saying, "Moving onto the economic conditions" as the first words of a sentence, with the voice being raised.

Example of structuring

If, for example, a surveyor was required to undertake a 10-minute business presentation to a group of 20 colleagues on why business skills are considered an important part of the firm's and its individual surveyor's work, initial comments by the presenter may comprise a welcome, a thank you to the audience for attending, an introduction to the presenter (although this could have been provided already by a chairman — see later) and stating the subject and purpose of the presentation. It may be possible to encapsulate the importance of business skills to the firm and its surveyors by providing initial illustrations, such as the importance of the firm's profile in the market, the role of business development functions, the need for a low-cost base to help ensure price competitiveness and the need for up-to-date economic and property market knowledge in view of the increasing number of investor clients being attracted to the firm. This can help an audience make immediate links between individual points they understand, but which they have perhaps not thought much about before in a co-ordinated way. Where members of the audience have different roles, the short overview needs to contain something for everyone, and this can extend to illustrating the team-based approach. The presenter could also briefly announce what the sections of the presentation will be, such as by stating the headings. There will be a final section of summary/conclusions, but the presenter may not include that in the list provided at the start.

The main sections of the talk can then follow, and in a way that flows from the initial brief overview. Therefore, as well as "Introduction" (covering the points above), sections could be "Raising the firm's profile"; "Business development initiatives"; "Pricing and profitability"; "Economic and property market updates"; and "The way forward". The last section is really summary and conclusions, but with a better title, and which while briefly reiterating key points, stops short of being repetitive. If the talk is accompanied by a PowerPoint presentation, key points could be listed under each section. Alternatively, the presenter may have an opening introductory slide, but make minimal use of PowerPoint by leaving only the main section headings for the audience to view.

Talk for the audience

As with reports where it is important to "write for the reader", a presenter needs to know the nature of the audience, and the information they will need in order to get the most out of the presentation. With both reports and presentations, the author/presenter can inadvertently assume knowledge, and omit information that is essential for the audience to gain a proper understanding. Surveyors may, for example, have experienced university lectures that begin at a level that is too advanced, or presentations from lawyers that similarly see comments "go over the heads" of the audience. The type of audience influences the language used (including industry jargon) as well as the content.

It is also important not to overload an audience with so much factual information that it is difficult to absorb. Again though, the content depends on the nature of the audience, and whether information is largely illustrative and not essential to remember, or whether information is essential to absorb pursuant to subsequent action. Handouts for the audience are a good way to introduce technically detailed information.

Notes for the audience

If notes are provided to the audience prior to a presentation, the presenter can be distracted by the actions and noise of the audience turning the pages as the presentation progresses. It can also mean that the audience spends more time focused on paperwork than being captivated by the presenter's comments. This includes the audience reading parts of the notes other than the areas currently being presented. An audience can also be less attentive if they feel that everything is contained in the notes anyway. On other occasions though, it will be important that the audience can make notes as the presentation progresses. It may work well for an outline of the talk to be provided to delegates at the beginning, with a more detailed set of notes being handed out at the end.

If notes are provided, and with reference to the comments above on structuring the presentation into sections, it is important that each section starts on a new page. If not suitably organised, the aggregate noise of paper being turned will momentarily drown out the speaker. The need to turn pages at suitable points (including mid-section) also provides scope for the presenter to take a pause. It is important that related points, which need to be presented without a pause, do not straddle pages. If necessary, notes can be included at the bottom of one page of the speaker's notes that remind them of what is on the next page, thus ensuring greater overall fluency.

If PowerPoint is used, the notes may comprise the same slides with space to add notes. All notes should be designed to correspond with the comments that the presenter will make at the opening of sections or the addressing of points within. If not, the audience can struggle to find their place in the notes, with the turning of pages again creating noise and distraction. Pages should be numbered so that a presenter can give the audience directions to the place in notes if need be.

The presenter's notes

During the presentation, the presenter will need a prompt regarding the points to cover (rather than simply reading from a full script). The type of presentation being delivered will influence the format to use and, as mentioned above, a business pitch requiring natural, largely unprompted delivery will be very different from a talk to over 100 delegates at a conference.

With reference to the points made in "Notes for the audience", the presenter's notes can be different to those provided to the audience, especially as more detail is likely to be required. The presenter still needs to be aware of where the audience will be in their notes, and directions may be added to the presenter's text, or pages may need to be renumbered in order to match the notes used by the audience.

In order to ensure a natural delivery, it is preferable to work from bullet points and other key notes, rather than a full script. It is important though that the presenter is familiar with the issues to be covered, and does, for example, look down at the notes during the presentation and not forget what needs to be said. A presenter also needs to ensure that reliance on relatively limited notes does not lead to undue creativity during the presentation, with the presenter saying more than was intended. This could cause the presentation to run over time, and/or the presentation having to be brought hurriedly to an end.

A full script can better keep a presenter on track, but risks sounding read out. However, some experienced presenters can draw on a full text but still sound as if comments are natural. As a combination between a full text and bullet points, words could be highlighted in bold, colour or increased font size within the text, or bullets could be set out, for example, down the left-hand third of the page, with the full text on the right. This enables the presenter to work from bullet points, but in the confidence that a full text is available if need be. A full text needs to be well spaced out, and it can be helpful to start a new line with each sentence — thus ensuring that the presenter can quickly find his place in the notes, and therefore make as much eye contact with the audience as possible.

A presenter needs to practise, and establish the most suitable approach for himself. Sometimes, presenters prepare a full script initially, with a view to actually working from bullet points on the day. This enables the precise content to be fixed, as opposed to being more indeterminate with bullet points, and changing slightly on each occasion the surveyor practises the presentation. A full text enables a draft to be circulated to others for observation and input, whereas bullet points are less helpful. Presenters also need to consider whether to work with notes on A4-sized paper, or use cards (bearing in mind that cards are neater and A4 sheets are more likely to shake through nervousness).

Despite the above comments, experienced presenters are able to take a reasonably flexible approach to delivery and on their reliance on notes — particularly as they develop the confidence to talk relatively spontaneously and know, through experience, that they will not be caught short of things to say. They keep an eye on the time being taken, and are able to enlarge on some points if need be, while also being able to omit certain points if necessary. However, if a handout is provided, the impression should not be given that sections/points are being left out.

Positioning of the presenter

Presentations are undertaken in different ways, including sitting at a table, standing behind a lectern or simply standing in front an audience. Standing up is most common, but a speaker may, for example, sit at a table on a stage and be sufficiently prominent to the audience. It would also be easier to sit down if the audience was in tiered seating.

A lectern enables the presenter to easily draw on notes and appear relatively composed. It also enables the hands to be placed on the lectern if need be, whereas standing directly in front of an audience can feel comfortable. Holding paper and a pen can work well, and be preferable to hands clasped behind the speaker's back or clasped facing the audience. Hands in pockets can look slovenly, but may help informality, at least with one hand. Very occasionally, a required informality could be achieved by sitting on a table.

Presenters always need to be conscious of the body language conveyed to the audience, and ensure that eye contact is maximised. Looking down or away from an audience needs to be avoided. Presenters should smile occasionally, although this can be difficult if not relaxed. A few subtle points of humour can, however, show that the presenter has a human side, rather than only an intense focus on facts. A suitably varied and inspiring tone of voice also helps to keep the audience's attention.

If a presenter works at a table and faces the audience, a tablecloth is ideally draped over the front. In the early stages of a presentation particularly, a presenter's legs can move around in an agitated way, and if there is no tablecloth, the presenter needs to consciously remain still.

Presenters could also sit down where there are small groups, which could be positioned round the same table. This enables eye lines between the presenter and the audience to be on the same level, rather than a presenter being positioned domineeringly over a group.

The positioning of the presenter may also have to interrelate with a PowerPoint or other display. It can come across poorly to the room when a presenter regularly faces the PowerPoint instead of the audience. If a microphone facility is not in place, the presenter's voice will fade for those at the back of a room. With reference to the section on "The presenter's notes", the presenter ideally sees a screen in front of him (such as a laptop or built-in facility) containing the PowerPoint information, and therefore remains able to face and address the audience. This will also help the presentation appear to be delivered more naturally. The presenter is, however, more easily able to face the PowerPoint screen display when pointing to pictures or plans, for example.

Use of PowerPoint

The use of PowerPoint is common in presentations, but is this always the best approach, and how should the PowerPoint presentation be structured? Presenters can sometimes give the impression of hiding behind PowerPoint, and so lose their ability to communicate as naturally as possible with the audience. A continually revolving PowerPoint presentation is too much for an audience to keep up with, and sometimes also too much for the presenter to manage. Also, the audience can be focused primarily on the screen, rather than facing the presenter. This means that the presenter does not best captivate the attention of the audience (and the audience needs to feel that they have attended a presentation, not simply a list of points). It is also difficult for an audience to read one thing while listening to another. As such, some presenters prefer to dispense with PowerPoint altogether, and address an audience more head-on. Indeed, as a general business point, an absolute orientation to substance and a deliberate suppression of style can come across well to certain client groups.

Within the presentation, PowerPoint can be used effectively to introduce points, with the audience's attention then returning to the presenter. This ensures the dual benefit of the impressive use of technology (including stylish graphics and designs) and the presenter showing his calibre. An impression of style above substance needs to be avoided, and the PowerPoint show should not be over the top. The amount of information placed on screen will also depend on the extent of notes being provided separately to the audience. It can be frustrating for an audience to be unable to copy notes from a PowerPoint presentation before the presenter moves on to the next slide.

PowerPoint is a good tool in projecting details of a host, sponsor, speaker and/ or firm at the start of a presentation. Similarly, at the end of a presentation, details of the speaker and firm can remain, including contact details, which may be a more subtle promotional method than a presenter announcing that he can be contacted. (Although the purpose of some presentations will be for promotional benefits, it cannot be seen by the audience as a sales pitch, and softer techniques are needed.)

As everyone is aware, technology can fail, creating a poor impression of the speaker and a presentation, even though faults may be beyond their control. The more ambitious the format, the greater potential embarrassment if things go wrong. Technology can be an accident waiting to happen, and it is imperative that PowerPoint or any other presentational aids work smoothly. Presenters must also have a contingency for technology breaking down, and be able to continue fluently from notes if need be, with minimal discomfort in adjusting. Arriving early at the event allows time for all logistics to be finalised, noting that staff at a venue might not be readily available.

Dealing with difficulties

Despite being experienced at delivering presentations, speakers can sometimes still freeze momentarily, and simply not know what needs to be said next.

The availability of water is a good outlet in the event of such difficulties. The presenter just needs to take a sip of water, and gather his thoughts, also giving him time to look down at his notes. To the audience, having looked up, and all eyes being on the presenter following a pause of unnatural length, a sip of water looks innocuous, and the audience is unconcerned.

If need be, the presenter may say that he has just thought of something else to tell the audience later, accompanied by the writing of a few notes. It may even be possible for a presenter to say, in a relaxed way, that he has forgotten what he was about to say, and will come back to it.

Late arrivals and people leaving for the toilet can be distracting, and there may be members of the audience talking to each other during the presentation. In such a situation, it may be necessary for a polite request to be made for them to cease, with the surrounding audience also welcoming this, and the offenders usually being contrite in their compliance. Very occasionally, a member of the audience might interrupt the presentation to ask a question, and can this be a surprise to the presenter. However, some presenters advise the audience at the start that they may interrupt and ask questions. A presenter may also observe a member of the audience smiling overtly, or even laughing for no apparent reason. Smiling is likely to be the individual's unconscious way of appreciating the comments being made by the presenter, and laughter is likely to be something said by someone else in the audience — neither of which are a poor reflection on the presenter's performance.

Questions from the audience

Questions from the audience can sometimes put a presenter on the spot. Presenters can appear accomplished from their presentation, but then struggle to deal with questions.

It may be possible to structure a presentation/event in a way that does not allow time for questions, but often a series of questions are expected by the audience. This can be a lively part of a presentation, raising interesting issues and enabling the presenter to further demonstrate his expertise.

When questions are being asked, and which appear to be lengthy, or have two or more parts to the question, it can be helpful for presenters to write down points. Parts of an answer can begin to be structured as the question is still being asked, and the response can be particularly articulate, with a presenter dealing with points one-by-one — as opposed to having to return to the person asking the question for a reminder of some of the points raised. Asking a question back to the question setter in order to gain more information may also be helpful. Thanking the person for asking the question, or commenting that it was a good question can give much-needed thinking time.

As well as inviting questions from the room, it is helpful if people are on hand to ask questions if need be, such as the chairman of an event, or the presenter's colleagues. Sometimes there are no questions from the room — not necessarily because there is nothing that anyone wishes to know, but because people can be reluctant to speak. Where there is a particular need for stage management, colleagues, for example, can be primed to ask certain questions, and even engage in a debate with the presenter. It is important, however, that the questions do not appear to the audience as being staged.

Role of a chairman

On occasions, the role of a presenter will be in the capacity of a chairman, such as to introduce an event and a speaker/s at the beginning, or to provide links between speakers/subjects throughout an event. The chairman may also co-ordinate questions from the audience, provide a vote of thanks and generally close the event.

A chairman may be confident at speaking to groups of people, but should remember that the invited speaker/s are the key people involved. Some chairmen have too much to say, delivered with indulgence and confidence, and leave the opening speaker starting in a shadow. A good chairman welcomes the audience, thanks the hosts and/or sponsors and provides introductions to presenters that help warm the audience to them. Presenters should supply the chairman with information about themselves that can be covered in an introduction. Some presenters, however, require only the very briefest of introductions in order that they can introduce themselves and their firm in their own way. For example, four surveyors from the same firm may have an opening speaker who introduces the team before delivering the first presentation personally, and then introduces the other speakers as their turn comes round. Alternatively, the first three presenters may introduce each other, with the fourth closing the session. The introductions by the surveyors themselves are likely to be more fluent and detailed than a chairman who has only just met the presenters and is working from quickly hand-written notes. It is still nevertheless favourable for the speakers and their firm when an independent host comments, for example, that a particular presenter is regarded as one of the leading surveyors in his field, as it would not be appropriate for the surveyor to announce this personally.

A chairman with too much to say may also overlap with the comments to be made by the speaker, and leave the speaker having to adjust within a presentation. It can be particularly difficult for a presenter when a chairman's remarks warrant a variation to the early comments that were planned. Similarly, if there is more than one speaker, each should know the others' remit and the areas being covered, and this will avoid overlaps.

Managing events and other issues

In order to create the right professional impression with a presentation, the presenter or colleagues need to be alert to surrounding logistics, or any other factors.

Consideration needs to be given to the way in which delegates will register for an event. The majority of people may arrive at the same time, with the need to provide delegate badges, for example, taking time and delaying the start of the presentation. On other occasions, such as where people have further to travel, arrivals and registrations may be spread over 30 minutes or more. Event confirmations, notifications of venues etc can highlight the need to arrive from, say, 10.45am for an 11am start. Notification of the availability of refreshments prior to the event can also help, with delegates registering on arrival, rather than on entry to the presentation itself.

While some venue providers are proficient at ensuring that the event will run smoothly, others may not be alert to the items requiring attention. Below are some useful checkpoints to help the event run smoothly:

- If the temperature seems high when the room is empty, it will be even hotter when the room is full. An audience remains more alert in a cool environment and the impression of disinterest to the presenters is also avoided. Windows may need to be fully opened for some time prior to an event, but closed because of outside noise just before the start.
- The lighting needs to be right. The audience does not, for example, want bright sunlight to unsettle them, nor too much light making a PowerPoint presentation hard to view.
- Speakers will need a glass of water. Tea and coffee arrangements need to be checked to see that milk, sugar and spoons are all in place. Paper cups may present the wrong image, and the kitchen staff need to be instructed to provide cups.
- Any telephones in the room should be unplugged. Evening events can clash with cleaners' hovering in corridors, and arrangements made to prevent this.
- Microphones need to be working well, and speakers rigged up rather than handing microphones across. Equipment may pick up the presenters' discussions amongst themselves, and the presenters will need to be notified of this.
- Presenters should not ask the audience if everyone can hear at the back as this gives the impression of lack of preparation. Presenters should speak with sufficient volume, clearly and without mumbling. When drawing on notes, it is preferable to dip the eyes down, rather than move the head down, and perhaps therefore lessen voice projection.
- Temporary directional signs may be needed within a building and on the door of the room itself (which should be printed rather than appearing handwritten on the day).
- The venue may have a board at its entrance advising of the name of the event/firm and the room in which the event is held, and this needs to be checked.
- Delegate badges may be needed.

Prior to the day, the venue needs to be aware of the number of chairs that are required, the layout of the room, and the arrangements, for example, regarding a top table (number of seats, access to a lectern). If there are 50 delegates in attendance, it would be better for the room to appear full, and the event successful, than would be the appearance if 50 people were spread thinly over 100 seats. Choice of room, the number of seats placed in the room, and the use of partitioning can help achieve this. There can also be a tendency with events that the younger the audience, the more inclined delegates are to fill up the back seats, thus leaving a void between the presenter and the audience if there are too many seats in the room. A tighter group of people also enables the presenter to better captivate the attention of the audience.

Consideration needs to be given to the location of access points, and the scope for late arrivals to enter the room discreetly, without disrupting the presenter and the audience. It can be particularly difficult for a presenter to concentrate when the audience become unsettled by a late arrival finding a seat, and some of the audience may momentarily miss parts of the talk. It can work well for late arrivals to be held back for, say, 10 minutes, and enter the room together after an assistant has caught the eye of the presenter who then invites in the delegates. The reservation of seats at the front or back of the room (as appropriate, depending on entry points) ensures that late arrivals will not be scrambling past people to get to seats. Effective planning at the venue and suitable notifications to delegates, such as regarding car parking, or possible traffic problems in the area, will avoid a large number of late arrivals.

Hosts/presenters should also be aware of arrangements regarding fire exits, for example, and be generally alert to health and safety and any other requirements. There may also be a limit on delegate numbers, and the squeezing in of a few extra people on chairs brought into the room, rather than designated seats, may not be possible. The host/sponsor of a venue might require formal announcements to be made. Initial announcements should also include a reminder for delegates to switch off mobile phones. Directions to toilets should be clear, and the facilities generally able to accommodate requirements being concentrated at breaks in the event. Similarly, the distribution of teas and coffees to large numbers at breaks needs to be handled with speed. Heavy lunches, as well as alcohol, can leave an audience feeling lethargic and looking disinterested. If a presenter has too much to eat, this can affect the sharpness of the presentation.

An audience may also appear disinterested and restless during a presentation at the end of the day following several hours of previous presentations. Therefore, the presenter has to make an extra effort to captivate the audience's attention. In addition to the points above, the audience could be asked to give a show of hands to early questions — although the presenter should not do this in a way that immediately puts the focus of attention on the audience as a way of hiding apprehension of dealing directly with the audience at the start of the presentation. Occasionally, raising the tone of voice and being particularly expressive also brings back the attention of an audience.

As another example of the finer points of presentation technique and event management, the notification of a break needs to be a final comment. If not handled properly, the audience can begin putting away their papers in anticipation, with the aggregate noise preventing anything further from being heard.

If surveyors need to make presentations to community groups regarding planning issues, or to residential tenants in respect of service charge management, heckling from the audience may be experienced. The presenter has to judge whether to deal with the issue immediately or, as may be preferable, advise the speaker that the presenter will come back to them later to consider their point. The potential for hecklers to be subtly put on the spot may be a way of restraining other hecklers.

A presenter also needs to consider what to wear and how his appearance is likely to come across to the audience.

Training techniques

Surveyors requiring further support in respect of presentation techniques can attend a course run by a specialist provider, arrange initiatives internally and/or in groups with other surveyors/friends, or work on the development of some skills personally.

Specialist courses will usually enable attendees to put the presentation guidance into practice, and cover the more detailed aspects of presentation technique in relation to presentations that attendees are required to prepare. Details are available from *businessCPD@mptcentre.org*.

Group work could involve presentations on current surveying issues, or case updates for colleagues, and the group could give feedback on presentation style, and how the presentation came across, and where there could be scope for improvement. A video camera could be positioned in the room for the group to review, or just for the surveyor to take away. This can help the presenter's unconscious poor habits to be revealed, and eradicated.

A surveyor looking to develop skills in speaking fluently, without over-reliance on notes, can set subject headings on which they then talk for one minute initially perhaps, and then increase the time. Talks of five and then 10 minutes could then be structured. Such practice helps train the mind. One early noticeable point will be the reduced numbers of "err", "umm" and the overuse of words such as "obviously", "actually", "basically", which tend to be included to give thinking time — all characteristics of a bad presentation that should be avoided. In covering non-technical subjects, such as "the route taken to work" or "update on the family" and technical subjects, such as "the impact of landlord and tenant changes", the difference in fluency should be seen — reiterating the above points about knowledge of subject matter, having thought about issues prior to the presentation and practice all contributing to good performance.

Special training may be needed in respect of media appearances although, for surveyors involved only occasionally with television and radio features, the production company will usually provide guidance on the day, facilitate practice runs, and repeat any elements that have not come across well. Typically, comments will need to be incisive, and key points sometimes made in a matter of seconds. Preparation is therefore essential, while ensuring that the comments come across as natural responses/comments, rather than rehearsed statements. As with the comments above, it can take time to develop the necessary fluency, particularly when being expected to perform in a new environment. The study of public relations may also be of interest to some surveyors.

Accelerating Personal Development

Success in business derives from a combination of natural ability and instinct, practical experience, making concerted observations of how other people and businesses operate, and education and training activity. Hard work and luck also play a part.

This chapter shows how surveyors' individual personal development can be accelerated. It begins by considering the forms of training and education activity that develops the skills necessary to embrace clients' business issues in property dealings, as well as keeping up to date with changes in market practice, legislation and such like. Although *Business Skills for General Practice Surveyors* relates to surveyors' case work in practice, a brief overview is still provided of the issues arising in the operational running of property businesses, ie "business management". The chapter then summarises the primary personal qualities typically seen in people who become highly successful, and how surveyors can hopefully emulate such characteristics.

Training and education

RICS promotes the theme of "lifelong learning", which means that at all stages of their careers, surveyors should evaluate their personal and business development aspirations on an ongoing basis, and plan suitable learning activities. The term Continuing Professional Development (CPD) is still used, and is broadly regarded as training activity.

Lifelong learning/CPD applies to surveying subjects in order to enhance technical and applied proficiency in practice, and to business and management subjects in order to command increasingly rewarding management positions. In between these areas, study of business issues further helps enhance the quality of service provided to clients/employers, as has been demonstrated in earlier chapters. For those surveyors making a considerable commitment to lifelong learning, the acceleration to personal development can be considerable.

RICS and business skills

The emphasis on business skills is reflected in the following commentary, which was drawn on by RICS in its lifelong learning promotional material:

For some businesses, lifelong learning will simply involve remaining up-to-date with changes in practice and legislation in order to ensure that a relatively static client base is serviced. Indeed, among the many small surveying practices in the UK and also abroad, many undertake work in similar areas, year-on-year.

For other businesses however, lifelong learning extends to strategic growth, achieved through the development of new areas of expertise. Here, lifelong learning incorporates the development of effective marketing and business development techniques, including improvements in profile which help secure better quality work, more prestigious clients and higher financial rewards. Business development can, in fact, be facilitated through training by the delivery of seminars to prospective clients, and other surveyors, and also, for example, through contributions to professional journals.

All businesses must keep up to date with changes in economic and property market conditions and the impact this can have on their clients and their own business. Factors such as developments in technology also affect the competitiveness and viability of existing activities, and can facilitate profitable opportunities for those remaining most up to date.

Qualifying learning activities

The type of activities that count as lifelong learning are summarised below. This is an extract from *Estates Gazette*'s Mainly for Students feature of 16 September 1999, of which Margaret Harris of the RICS education department was joint author.

The most effective training programmes are those which reflect the nature of an organisation and the roles of individual personnel. A local authority or other in-house property team, for example, may see individual surveyors covering rent reviews, lease renewals and other landlord and tenant cases, as well as sales and lettings work, together with involvement in areas such as rating or asset valuation.

Surveyors in smaller private practices may cover a similar range of work, but surveyors with the larger firms may have relatively specialised roles — and therefore have very different training requirements.

Most organisations possess a level of expertise among their surveyors which can help develop the skills of others without the expense of formal training events. Surveyors' involvement in training could range from the day-to-day support provided for junior staff to the devising and delivery of training for qualified colleagues. Clients could be invited to talks, as could surveyors from other firms.

A company's solicitors could be encouraged to deliver seminars. Many already see the commercial benefit of doing so, and surveying practices could do more in this respect.

The organisation's provision of refreshments can be an effective way of ensuring that some training events take place outside fee-earning hours, with surrounding social activity adding interest.

Training sessions conducted on a suitable group basis can also enable team-building skills and morale to be improved. This includes support staff and younger surveyors, for example, being seen as a more valued part of the team. Indeed, the ability for support staff to deal with basic property cases, following training, can present cost and resource flexibility to an organisation.

Softer subjects, such as negotiation, report writing and coping with change, are also important, but tend to be given lesser priority than topics that are more fee related.

When work pressures tend to take priority over the arranging of training activities, it is important that the organisation designates personnel with specific responsibility for ensuring that training does take place.

Structured training

To most qualified surveyors, structured training means much more than undertaking 20 hours per year of continuing professional development in accordance with the RICS requirements. 'Structured' means that training activities stem from pre-identified requirements as part of an ongoing programme, as opposed to occurring on an ad hoc basis.

For individual surveyors, structured training involves identifying training requirements through the RICS Personal Development Planner (now part of lifelong learning and CPD on-line recording). Attention is focused on strengths and weaknesses, and on how training can aid personal development, along with business opportunities.

For organisations, structured training involves ensuring that surveyors, and also support staff, enjoy the best opportunity to develop their skills.

Private study

RICS CPD works on the basis of self-assessment. Surveyors determine the suitability of the training initiatives themselves, and record the activities undertaken, together with the number of hours devoted.

Structured private study can qualify for up to two-thirds of RICS minimum CPD requirement. This can be a particularly effective form of study, especially as surveyors have a differing knowledge and training requirements. Private study should generally be structured around specific subjects, or be part of wider learning initiatives. Simply reading *Estates Gazette*, would not, for example, be satisfactory, but the study of particular articles therein may be.

Articles do not require a 'CPD' or 'distance learning' type of designation, or the need to obtain certification from the provider, to count as CPD. When providers state the number of hours available, this must be regarded as indicative, with the recorded number of hours being determined by surveyors themselves. Private study could also comprise preliminary research, and/or subsequent study around a course or seminar.

Surveyors attending courses should disseminate the information to colleagues, and can also arrange discussion groups. These can consider the various issues in relation to the work in which the organisation's surveyors are usually involved.

Training events and training providers

Courses and seminars are particularly useful for specialist requirements, and also where it is necessary to be informed of new legislation, case law, market practice and so on. Many of these courses are useful because they cover many topics within a single day. The quality, content and suitability of training events can sometimes be difficult to determine in advance of attendance, especially with a large number of training providers in the market. Most commercial property training companies are in business because of an established demand for the quality of training they offer.

Qualified surveyors undertake a minimum of 60 hours' lifelong learning/CPD over any three-year period, equating to 20 hours per year, with a minimum of 10 hours being required each year. APC candidates undertake 48 hours of "professional development" (which is the equivalent to qualified surveyors' lifelong learning/CPD). However, the RICS highlights the importance of the quality of learning, with study areas reflecting surveyors' personal development needs, and supporting the work they undertake in practice. Since January 2004, on becoming MRICS qualified, surveyors must record their lifelong learning/CPD via the RICS website.

Surveyors considering taking formal business qualifications, including a Masters in Business Administration (MBA) should liaise with the RICS education staff.

Structured training in practice

As part of the RICS/Advantage West Midlands Annual Property Education Debate in 2003, a number of firms summarised their approach to training. Set out below is the RICS commentary on the structured training programmes implemented at GVA Grimley by Scott Kind, its UK training and development manager. Examples are then provided of how training functions can interrelate with business development opportunities.

> GVA Grimley's training structure reflects its position as one of the UK's largest commercial property consultancies. It includes:
>
> - In-house induction to fee earners, graduates and support staff
> - In-house IT training
> - In-house business and financial training
> - Graduate training, in particular the RICS Assessment of Professional Competence (APC)
> - RICS surveyors' and RTPI planners' CPD training, and similar support to other professionals, such as an in-house accountant
> - Support to all staff in respect of taking formal qualifications, ranging from NVQs to masters' degrees.
>
> CPD training in subject areas takes place by a combination of in-house events, attendance at external courses and private study material. These cover specialist updates in line with the need to provide leading advice to clients, and general updates on wider industry issues.
>
> In-house events include the delivery of seminars by GVA Grimley personnel to colleagues. Surveyors/planners are also involved in the provision of CPD for others by speaking at external events.
>
> GVA Grimley's training team, based in Birmingham, co-ordinates the overall programme, liaising with senior personnel in the business units, and appointed training co-ordinators around the country. The programme is supported by a monthly training newsletter, and a central library system, with smaller resource centres being operated on an office/departmental basis as required.
>
> Scott Kind, GVA Grimley's training and development manager, sees structured APC training as an important platform for continued learning after professional qualification.

He reports that: "The company's graduate training programme provides breadth of experience in a number of departments, and also a grounding in business skills. On qualification, the focus turns to the development of business, management and financial skills, and also the further development of subject-based and market knowledge. At this stage in surveyors' careers in a large private practice, this usually relates to the field of work in which staff wish to build a long-term career. A fast-track option has been established for staff identified during the APC training programme as being of the technical and managerial calibre to take the business forward, and aspire to partner status within 10 years of MRICS qualification. Here, staff successfully completing their APC will commence a post-graduate business qualification immediately. In fact, some may have begun this prior to finalising their APC training period. GVA Grimley ensures that outputs from some of the training initiatives are available to all surveyors, in particular to APC candidates. Training support is also available to key clients as part of GVA Grimley's overall business development initiatives. The training team is able to provide advice to clients with property teams on the structuring of their own training initiatives."

Training and business development

As well as training being an important function in ensuring that surveyors are up to date with market changes, it can interrelate with business development initiatives. An example is set out below of a summary of Part III of the Disability Discrimination Act 1995 (effective from October 2004). It combines a training-related overview, training policy and a summary of business development opportunities (albeit not extensive for the practice concerned). The firm's name is replaced with "the practice".

The Act involves 'service providers' ensuring that there are no physical barriers to disabled users. Examples of service providers include retail and leisure facilities, and the provision of public services by local authorities. A critical distinction is that Part III covers access to services, not buildings (although for ease of explanation, reference is made below to buildings being 'DDA compliant'). Office occupiers such as (the practice), where access is gained only by staff and clients, are not service providers. Some non-service providers nevertheless seek fully compliant space, such as when letting or purchasing. Together with the demand of service providers for compliant space, this can affect relative demand, rental levels and investment returns for compliant, as against non-compliant, premises.

Other elements of the legislation cover new and refurbished buildings, and employment related provisions, such as the requirement to ensure that any disabled employees can carry out their tasks without difficulty.

Assessment of opportunities
The practice has considered that DDA does not present a major opportunity to win instructions from new clients solely for access audits, and the administering of any consequent works. The practice will not therefore be specifically profiled in the market to provide such a service — although DDA related advice will still be listed in the usual range of services available in general promotional material. Also, although not actively seeking instructions for access audits from new clients, the practice would still take on any such requests, provided fee levels were adequate.

This approach reflects the number of other consultants looking to provide services in this expanding market, including smaller ventures working for relatively low fees on what is relatively straightforward work. Also, such consultants tend to be in the building surveying/construction side of the industry, rather than in the practice's more mainstream general practice functions.

Servicing existing clients

It is important that existing clients' requirements in respect of DDA are serviced effectively. This includes property management functions in particular, acting for both retained landlords and tenants. Here, a familiarity is needed of landlord and tenant issues, including special provisions under the Act, and of issues such as responsibility for common areas, and recoverability of expenditure through service charges.

Wider management consultancy opportunities

Where service providers offer services within a number of parts of their portfolio, a full strategic review of their property holdings may be necessary to ensure that compliance is most cost effective — ie solutions may involve concentrating occupation within parts of a building, and/or reducing occupation to fewer buildings. In some cases, therefore, DDA compliance may provide a lead for wider consultancy work.

Training support

Training support in respect of DDA broadly depends on the requirements of each surveying function. Building surveyors are attending external courses and arranging internal workshops, drawing also on a range of library resources in respect of DDA. This includes keeping up to date with changes in legislation, building regulations, etc. Property management surveyors and building surveyors will be involved in some joint seminars, including to consider any cross-selling opportunities between departments. Management consultancy surveyors may be involved in similar groups, depending on the work in which they are involved.

Other surveyors will be kept up to date on the implications on their work through the provision of concise study material, and lunchtime or evening discussion sessions led by managers. Issues include, for example, valuers' identification of non-compliant premises, any affect on valuation and the use of disclaimers — or landlord and tenant surveyors' assessment of rental adjustments between compliant and non-compliant buildings, and how to deal with factors such as useable space lost due to works necessary to meet DDA requirements. All surveyors will, of course, see DDA impact on their work to some degree. Much of the training material and other material will be posted on the intranet, with key updates being disseminated to surveyors via training contact points (see later) in each department.

Business development support

In order to support fee earners in any approaches being made to clients, a client DDA bulletin is being prepared. This concentrates on the wider advisory aspects relating to DDA, particularly for larger clients, rather than seeking to win individual audit instructions from new clients. It is, of course, important that within business pitches, especially to government clients, the practice remains very conscious of the importance of legislation such as DDA.

Training contact points

The training team will be establishing contact points in each department in respect of the implementation of DDA training for all fee earners. This will comprise a single contact where departments wish to thereafter facilitate updates for their team, or alternatively, contact points can be provided within each office for each department. The training team will also be drawing on (name excluded) of the London office as a national DDA training contact in view of his specialist work in the field. This will ensure that the practice is equipped with the latest information affecting the work of fee earners.

Management consultancy

The following comment from RICS Management Consultancy faculty also shows how education and training can be part of a promotional package, and how the faculty, effectively as a business with RICS, determines a structured approach.

> The faculty will provide to its market place through its members world-class advice for strategy building, tactical planning and the highest quality consultancy skills to implement these tasks (property or general business). Our market place and our customers require us to be better trained in economics, business and management skills, and to communicate our unique technical skills for the benefit of our customers. Our strong education, APC and CPD culture and academic programme will demand more of our members in terms of ability, and lifelong learning. This will contribute to the rising status of the management consultancy chartered surveyor in the worlds of commerce, government, property, construction and education. This will be the flagship faculty and the property advisor of first choice.
>
> The faculty has adopted the following aims and objectives relating to education and training:

- To 'educate' members of RICS in understanding the benefits that management consultancy techniques can bring and the contribution that they, as members of the Institution, can make in the wider business environment. In addition, the faculty is seeking to raise awareness of the importance of strong business management procedures in improving productivity and product development. The faculty will be using the results of its research programme to identify the skills essential to members wishing to develop their careers in this direction.
- To 'educate' the external market, including existing and future clients, other professional bodies, trade associations, consumer groups and the media, of the unique ability of RICS members to combine legal, business and commercial acumen with unrivalled property expertise.
- To promote the management consultancy APC as widely as possible, whilst ensuring that there is adequate business skill content in courses at undergraduate level, in particular, economics and strategy. In addition, that there are post-graduate courses, both full and part-time, to enable members to develop their careers and market themselves successfully against other more recognised business professionals.
- To ensure adequate provision of high quality, accessible CPD/professional guidance/market intelligence covering all aspects of the faculty's remit; coupled with leading-edge professional guidance/accreditation schemes to ensure members are equipped to compete in today's fiercely competitive market. In addition, the faculty is committed to developing its micro-site to provide members with 'one-stop shop' access to up-to-the-minute market and business trend intelligence.

The faculty is also looking to maximise the potential of networking, both at faculty and member level. The opportunities to be gained from working together with other organisations and sharing information and resources cannot be over-estimated. An inclusive attitude towards working will also extend to relations with other RICS faculties and forums. In addition, the faculty will be looking to further establish its profile by attending and participating in industry wide events and ensuring a presence at national and international business and management conferences and seminars.

As indicated above and in chapter 4, management consultancy work gives rise to the need for particular business skills. Some are outside the scope of *Business Skills for General Practice Surveyors*, but for details of further learning areas, the RICS Management Consultancy faculty can be contacted on Tel: 020 7222 7000.

Business management

All businesses have their own economics, dynamics and drivers that need to be understood by principals, and also embraced by staff. It will be important, for example, to regularly review the performance of a surveying consultancy, maintaining an entrepreneurial eye and looking to be creative in terms of new earnings opportunities.

Strategic issues

A business needs its strategic thinkers who are able to formulate growth plans. Strategic planning will reflect the identified opportunities and threats for the business, including those relating to economic, property market, surveying market and technological change. Senior personnel need to keep in touch with their markets at a grass roots level, particularly as they can easily become remote from day-to-day issues and finer market trends. Consideration similarly needs to be given to the scope for operational efficiencies, including cost savings.

Business development

Marketing and business development functions will be important in achieving business growth. For property consultancies, this cannot be a stand-alone function, and there are many contributing factors in consistently winning new instructions, growing year-on-year, and achieving high profitability. The right profile and stature is needed within the market at large, and regarding individual fee lines — including in the eyes of different client groups. The track record and expertise of the business and the individuals handling instructions are important, and other parts of the business may also have to perform as part of a composite instruction. Not to be forgotten is the high standard of attention afforded to existing clients, who themselves and through their contacts represent a means of business development.

Short and long-term issues

While short-term pressures placed on personnel can support current year profitability, the right culture regarding the medium and long-term stages of business growth needs to be embraced. Divisional structures, bonus systems, internal accounting issues and central overheads all need to strike the right balance in individuals performing for the wider business, and not only themselves and their localised interests.

Empowering others to perform

Property is a people business, and all personnel need to be empowered to perform — needing to be motivated, rewarded and retained, while minimising the cost base. This will include leadership skills for senior personnel, people management skills for a business' middle managers, and a combination of property and business skills for younger surveyors. The larger the business, the greater the need for senior personnel to empower others. Aspects such as vision, focus, strategy, entrepreneurial acumen, creativity, profile, competitiveness, cost base, profit/profit margins and earnings growth need to be engendered in others — including individuals lacking natural instincts in such areas. A business may be well established, and its founders' entrepreneurial instincts long-gone. Succession planning will also be important, with leaders of the future having to be identified within the business or attracted from elsewhere.

Total business management

"Total business management" means that, in order to optimise profitability, property businesses need to successfully co-ordinate key interrelated elements such as those highlighted above. Indeed, it is surveyors' integrated understanding of property markets, surveying functions, business and managerial issues, and marketing and business development activity that drives a business and its profitability forward.

Specialist marketing and business development advisers will have a role to play, but surveyors need to be actively involved in the marketing and business development processes, particularly as such advisers do not necessarily understand the economics, dynamics and drivers of the property industry and individual business. A low-earning surveying function for a particular property consultancy (such as property management) could be the lead for highly lucrative other business (such as agency: lettings and acquisitions) as well as providing defensive benefits against market/sector downturn. Likewise, a small and relatively unprofitable amount of agency business could help a surveying consultancy gain profile in the market, which helps more lucrative business to be secured (as well as market knowledge/intelligence derived through agency work being important in supporting other functions).

As another illustration of total business management, a low-cost base is important in regularly winning business on both price and profile, and in securing optimum profit margins. Similarly, accounting and taxation advantages will need exploiting, not only to enhance profitability, but also to facilitate price-competitiveness.

Psychology and relationships

Surveyors also need to appreciate commercial and interpersonal psychologies. Examples include the fickle way that business can sometimes be awarded to surveying firms by clients, where, aside from price and perceived potential to perform, decisions sometimes reflect whether clients/representatives like and will get on with the appointed surveyors. Friendships, family connections, other contacts, gifts, hospitality and certain mutually beneficial arrangements can all be of influence, and the best surveyor/firm does not necessarily win the instruction.

Surveyors need to consider how they can best endear themselves to clients and colleagues. The profession has its many personalities, ranging from the down to earth instantly likeable surveyor, to the surveyor seeking to present an image of success beyond actual reality through attire, grooming and their general manner, but which can quickly alienate others who are not of like tendency.

Relationships with some clients will be built through similar interest in local sports teams whereas in other situations, skills extend to the ability to discuss cultural, historical and topical, issues; express opinion on a wine list or possess other social skills. All surveyors should be familiar with matters such as their company's share price, major activities/schemes, roles of its various departments and recently reported financial results in order to present the right impression if clients raise such issues. Knowledge of the residential sector is also helpful, despite commercial property surveyors not being involved in the sector, as clients will usually have an interest, and not be impressed by advisers who appear unaware.

Others issues

There are a multitude of factors involved in business management and this chapter provides only an initial insight.

Human resources functions will include recruitment, training, coaching, performance appraisal/management, compliance with employment legislation, alertness to health and safety issues, managing redundancy programmes, managing change and handling absenteeism.

Senior personnel in all sectors will need to handle the media, as well as working with public relations advisers to present the right image of their organisation in the market.

Senior personnel may need to examine the merits of acquisitions, mergers, stock market flotation, or sales of parts of the business. The expansion or contraction of a business, or certain parts, will need to be managed effectively. International issues are increasingly relevant for the large property consultancies, as are issues such as corporate governance, environmental responsibility and equal opportunities.

Compliance with company law and accounting/taxation law will be important. A business may also be hindered by political issues, particularly in the public sector.

Further studies

For the purpose of surveyors' further study, and reflecting the application of *Business Skills for General Practice Surveyors* being mainly for younger surveyors, four broad areas can be considered:

- business management
- business development
- people management
- business growth.

Business management concerns the general operational running of the business, including its administrative/cost and surveying/earnings functions. Business development involves the winning of new clients and new instructions, and therefore enhancing earnings. People management is a means of getting the best out of people, and also the general management of relationships among people. Business growth involves evaluating the merits of stages of growth (including the many implications this might have) and managing growth effectively.

While all four elements would be relevant to a surveying business or a property company, a public sector or corporate sector property team would be concerned primarily with business management and people management. Here, the cost base and operational efficiencies will be key areas and, at more senior level, strategic thinking, leadership skills, people management and communication techniques will be important.

Observing others: people management

As well as study through a commitment to lifelong learning, including courses/seminars and private study, observations can be made in surveyors' workplace. People management provides an example, with surveyors learning from the qualities of effective managers and observing the contrast between good managers and bad managers. The following extract from training notes helps illustrate some of the issues, including the links between people management and business success.

Consider, for example, two managers (partners or directors) in a larger surveying practice. They are equally busy, but one thrives on all work seemingly being urgent, sub-consciously raising self-esteem through the urgency, pressure and the sense of importance that this is thought to bestow. There is little time to deal with the needs of colleagues, whether through case queries or general interest in their work. Support staff are treated poorly, and 'attitude' comes across to others — except clients, for which an instant reversal to charm is still assured. The manager becomes increasingly disliked, such as by those having to complete reports by an urgent deadline and work well into the evening, but which then sit in the manager's in-tray for several days subsequently. Time will not be available for business development activity. Success is assumed because profits are being made, and people are busy, but thought is not given to the growth of the business, and expansion opportunities being lost. Secretaries and other support staff seem to come and go, graduates depart soon after becoming qualified, and other surveyors move on more frequently than in other departments where better managers are in place.

In another department, a high quality manger works on the principle that 'nothing is urgent in a well-managed business'. Business is better planned, and the manager feels more socially and professionally secure — pro-actively seeking to encourage and get the best out of colleagues. Time is available for queries, and thought about business development sees the department's fee income and profitability gradually grow. Social activity is arranged, staff retention is high, petty days of sick leave are non-existent, and work is enjoyable — all adding to high productivity.

Technically, the two managers are likely to be equally able, and have similar fee earning capacity as individuals. However, one will be vital in taking the business forward, whereas the other's influence could be detrimental. The inherent lack of business skills in the profession (as stated by RICS) partly accounts for why poor managers hold management positions. It also accounts for why ordinary surveyors with managerial skills can hold high positions.

There are, of course, variations between the above examples. Commercially aggressive managers with little time for colleagues may still be held in high regard, such as because of the support they provide when colleagues run into difficulties, and the rewards facilitated for their team through pay rises, bonuses and other benefits. Similarly, the likeable, good people manager, may simply not make the most money for the business — another example of how the more belligerent, poor people manager, can often be the one who does make the most money. The example shows the importance of drawing on the right calibre management. It also shows, with respect to training and development, that even the most senior people in a business may benefit from training, including re-orientation of how their harder property and business skills can be complemented by better softer skills to further enhance performance. Teams also need the right diversity and blend of individuals' strengths.

Becoming brilliant

A heading such as "becoming brilliant" sounds akin to publications such as "how to become a millionaire", but surveyors can draw on the characteristics of highly successful people, and weave their own way to greater riches.

A professional services industry such as surveying may not seem to have the same scope to deploy entrepreneurial acumen and business creativity as other industries (such as retailing, leisure, manufacturing) but surveyors are still well placed to secure high financial rewards from the property sector. Brilliance could be achieved by exceptional performance within one of the large practices, being a principal within a successful small/mid-size niche practice, setting up a new surveying practice, or exploiting property investment and development interests.

"What makes a millionaire?" is a good question in considering how "becoming brilliant" might be defined. In defining millionaire status as a target, this excludes inherited wealth, increases in house prices, reaching millionaire status eventually (after, say, 50) and headstarts, such as having surveyors within the family, family working for clients of the profession and family businesses. Whole books are dedicated to such matters, but set out briefly below are initial points of illustration.

Attitude to personal development

A very simple distinction between highly successful people and others whose careers progress minimally over 20–30 years is the attitude to personal development. In the surveying profession, there are many surveyors who, after becoming professionally qualified, undertake only minimal further learning activity for the rest of their careers and generally view CPD only as a RICS compliance requirement. Modest pay increases and promotional opportunities are still often secured over the long term, but reflecting factors such as age and the retirement or departure of others. Such people do not strive, have no hunger (or even professional pride) and generally prefer to remain in the comfort zone.

There are other surveyors who are more aspirational, but whose application in achieving advancement is a different form of laziness. They are prepared to commit long hours by way of physical commitment, but always in an automated way regarding the same straightforward case work. Laziness lies in not searching out more complex case work, not being prepared to seek employment with another firm and not being willing to commit efforts to learning/CPD. The payback from lifelong learning/CPD is, of course, an ability to undertake case work with greater acumen and therefore proficiency, and command increasingly complex, more valuable case work. This further develops experience, expertise, track record, profile, fee-earning capacity and personal rewards. For this reason, the right attitude to personal development accelerates surveyors' performance, financial rewards and general success.

Brilliant people search out accelerators. This includes platforms for progress — ie identifying one thing that will then lead to another, which in turn leads to another, and then another (rather than contemplating only immediate possible benefits). It is also important not to shy away from opportunities to accelerate personal development, and instead be pro-active and bold.

Brilliant people also recognise that they do not know everything, and believe that there is always something more to learn, despite how accomplished they may actually be. People do not know what they do not know, and do not know what they have not seen or heard. Examples of the wrong attitude are senior/mid-managers who think that "there is not much you can tell me about ... that I don't know" (despite the achievements of others being far greater), or graduates who think they are a "bit special".

Entrepreneurial eye

An entrepreneurial eye is key. This includes thinking about things that others fail to think about, seeing things that others do not see, and doing things that others do not wish to do. Highly successful people see gaps in the market (including "golden gaps"), create submarkets and provide differentiated services. They understand psychologies and business drivers, and critically, understand the markets in which their ventures operate. They are conscious of profile, and how an ever-improving track record helps bigger and better business to be achieved. "Brilliant people" often keep as quiet as they can as they can about their success, and avoid opening the eyes of others to the sources of their success. Being in the right place at the right time, and the occasional element of luck, can also be important.

To become brilliant in delivering professional and consultancy services by setting up alone, risk need not be great, nor finance a major issue, but being alert to competition and other threats is essential. It is also important to judge the form of growth that is appropriate, and the general direction of business interests. Property investment and development can carry risk, but the ability to read markets, the patience to not invest at certain points in time, and the expertise to develop bolt-on revenue streams can help achieve high financial rewards.

Vision

"Pounds not pennies" means that while the majority of surveyors chase the short-term, day-to-day pennies, the more thoughtful, patient surveyor, finds ways to earn pounds, albeit beginning some time later once their carefully formulated plans come to fruition. Early career examples include sacrificing salary to undertake formal qualifications or CPD/learning in key areas, or taking a job that does not pay the most but provides a springboard to better things. Astute elements of business development include searching for opportunities that involve no initial remuneration but can lead to highly beneficial opportunities, not least because the majority do not search for such opportunities (ie their focus is "today's pennies" not "tomorrow's pounds"). The skill is also in knowing which forms of such investment tend to reap more distant returns — which, as with the variability in success of mainstream marketing and business development initiatives, can be difficult. Confidence, belief, vision, focus, motivation and determination are key ingredients.

Focus

Highly successful people typically focus on the success achievable in specifically defined areas, niche concepts, etc, rather than looking to span a variety of such areas. An investor specialising in the industrial sector, for example, may go against the investment maxim of not putting the all eggs into one basket and diversifying risk, but achieves higher financial returns through the expertise and brilliance deployed. A small surveying firm, or sole trader, offering too many services can appear to be a "jack of all trades" enterprise, lacking expertise in any area, whereas a firm/surveyor profiled as a specialist can be seen as having leading expertise, and may achieve higher fee levels and profits because of this.

It is also important to set personal goals — and then, when they are reached, set new goals.

Thinking time

Often, the highest quality analytical thoughts that lead to new opportunities and/ or the refinement to current activities come when relaxed. Therefore, it is important to create opportunities for thinking time. Breaks of three to four days can be particularly helpful, giving time for immediate thoughts about current work to disappear, allowing more strategic issues to be contemplated before a too long an absence from work nulls the mind to any inspiration as holiday surroundings take over. At work, brainstorming sessions with colleagues can also be helpful, as can small research initiatives. When senior managers prepare detailed end-year reports, if sufficient exclusive time is taken out, rather than reports being prepared piecemeal, the more intense and cohesive reflection of the year's trading and opportunities for the business can illicit more in-depth analysis and new ideas for the future.

Other examples of the importance of both focus and thinking time include a core element of a business suffering as takeovers, mergers and other expansion into non-core areas deflects the thoughts and attention of senior management. Professional sports people can see their performance suffer due to a loss of focus when becoming involved in commercial activities, and thinking time hitherto reserved for evaluating performance and rectifying deficiencies is now being devoted elsewhere.

Similarly, surveyors whose thinking time is prioritised towards matters such as social activities, family issues, personal image, the impressions of others, and irrelevancies, such as which mobile phone ring tone to buy, deny themselves the quantity and quality of thinking time.

Communication

Brilliant managers and brilliant surveyors are usually good communicators. Communication ranges from keeping employees informed on company issues to the tone of e-mails and memos. People whose approach to business includes ignoring certain letters and e-mails, and hoping certain issues go away, lack certain attributes in "becoming brilliant". This includes those who are uncomfortable addressing the negative — such as saying no or giving unfavourable news.

Successful people also tend to have time for others. The often-used phrase "have not got time", is really, of course, a less offensive way of dealing with the allocation of priorities. Where there is talent, there is time!

It is important to understand people, and generally be a good "people manager" in order to get the best from others, including staff, clients and contacts. Breakdowns in communication can alienate, frustrate and discourage others, including people with whom the preservation of relationships and support is important. The ability to understand market, consumer, client, and staff psychology, including fickle factors, emotional aspects and human behaviour, is also key. Good temperament and the ability to make rational decisions quickly, including under pressure, is important. Successful business people are also able to assess/detect the genuine capability of others — as opposed to simply making judgment based on personality and persona.

Instinct and judgment

Another important business point is the ability to make judgment when the advice of others is unavailable, or there is no guidance in print: some people have a natural ability to make the right judgments, whereas others instinctively see obstacles that avoid making the decision and taking action which may carry adverse consequences if wrong. As well as being instinctive, successful people "think things through" and "consider the consequences". This involves working through likely consequences of decisions, actions, policy etc, rather than resting at superficial notion that actions, policy, etc appear correct. This is why market research, consultation and simply running ideas past people are important.

Dealing with the dull

In business/surveying, often there are aspects of work that are less inspiring than others. Highly successful people tackle dull day-to-day and more complex work with the same energy as the more inspiring aspects of their business. They realise that it is integral to the success of their business. People who struggle to deal with difficult and/or dull tasks, opting for more interesting and exhilarating alternatives, will not become brilliant. The surveyor enjoying the comfortable routine of day-to-day case work, and wishing to avoid difficult negotiations, or the requirement to deliver a presentation will not become brilliant. Similarly, the type of person who shies away from progressing a year-end tax return until the last minute, or a graduate who always puts off study efforts, and puts back targets to complete written submissions for the APC lacks the psyche to become brilliant. "Bit-brains" are people who can deal highly effectively, for example, with a large number of quick and easy e-mail responses, but leave aside the reading or writing more extensive reports and more complex issues for another day (and then another day). Again, they will not become brilliant. Rare exceptions to the above examples are entrepreneurs who prefer to be relatively untroubled by paperwork. However, instead they ensure support from the right professional advisers and employees, and still understand all the facets of their business and their market, including the most mundane issues.

Commitment

Long working hours can help maintain focus and minimise distraction and, although some work-life balance is needed, absolute commitment and focus to a venture or initiative may be helpful over a certain period of time. Long hours also create a sense fulfilment, similar to running 10 miles being more rewarding than running two miles. This is one reason why some people work long hours. It is typically under the guise of work being urgent and necessarily having to be undertaken, but is often a false urgency subconsciously induced by other factors, including physical and mental fulfilment. This is what is meant by "workaholics" and "adrenalin junkies". For some people though, the motivation to work excessive hours is simply financial greed. Drawing on adrenalin is an important part of striving for brilliance, and can enable long hours to be worked, despite tiredness.

Other issues

There are a multitude of other factors, but the above at least provides a brief insight. Surveyors can also observe and analyse other people and circumstances, and see broad trends.

The type of person who complains, for example, is more likely to be unsuccessful and frustrated, than successful and content. The psychology is that the lowly individual needs to feel better about themselves, and achieves this through knocking others, self-delusion as to their high standards, calibre in comparison etc. Brilliant people are perennially positive (although not deluded or naïve as to the negative), and see obstacles as a challenge to overcome. Unsuccessful people bemoan obstacles and difficulties and also welcome them as reason for lack of action or failure to achieve.

Brilliant people strive to attend meetings on time, whereas their unsuccessful counterparts are indifferent as to such minor achievements, and are often late, even if only by a few minutes. Other characteristics seen more often with unsuccessful people include greater absenteeism from work and weak underlying reasons/illness, being the members of staff who cannot seem to make the journey to work when there has been a few inches of snow, easing down into Christmas festivities at the start of December, and viewing Friday afternoons as non-work time. Successful people thrive on change and the challenges it brings, whereas unsuccessful people are more likely to resist change because they lack the ability and/or confidence in adapting (or shy away from the harder work involved in adapting). Change for change sake though is, of course, potentially detrimental — such as a new manager or other member of staff making unnecessary or marginal changes in order to facilitate a more noticeable personal impact, and/or point of reference for further measurements of success.

Case illustrations

As an example of how surveyors can make the most of their potential by deploying some of the above facets, the following is extracted from *Starting and*

Developing a Surveying Business, and follows comments on issues affecting viability and profitability.

There are, of course, many factors to consider when determining the format for the business, but this does at least show how with a skilfully tailored concept, a busy and reasonably talented home-based self-employed surveyor can achieve strong profitability. As an example, a Midlands based surveyor in his early 30s, setting up alone from home with no secretarial support, and without existing contacts/clients in the late 1990s, reached annual profitability of £100,000 within the first full financial/tax year. A combination of property investment, property management and business management consultancy enabled high added value to be provided to growing small investment companies at affordable fees, with low costs being secured through home working and basic information technology skills which avoided the need for a secretary. The surveyor's equivalent salary if working for a property consultancy would not have exceeded £30,000. Today, the surveyor need not work again, although still enjoys a combination of property, advisory and educational interests. The principles embraced by the surveyor in setting up, and in business generally, are reflected in the issues emphasised in *Starting and Developing a Surveying Business*.

The additional extract below further illustrates some of the above qualities and characteristics in achieving success.

As a general point, a way for a self-employed surveyor or small practice to expand is to continually seek platforms for progress — identifying activities that will lead to other, better and more profitable, activities and being selective about taking on instructions of low value and/or which do not fit with the wider aspirations of the business. Such platforms may not necessarily be the most profitable work at the time, but it is their scope to lead to better things that makes them worthwhile. The Midlands based surveyor referred to in Chapter 1, for example, reached over £200,000 per annum in the third full financial/tax year, still working alone from home without staff. Although a rather extreme example, and representing an excessive number of hours being committed per week, it is an illustration of how concerted dedication in the short term can lead to a subsequent winding down to enjoy increased leisure time thereafter while still working on selective business interests. In contrast, surveyors/businesses simply content with being busy, do not look to develop further stages of growth. The range of services, location, markets/client base may, however, simply inhibit growth anyway for an individual surveyor/small business.

As a further illustration of the concept of simple contentment from which individuals and businesses can suffer, the following extract refers to the primary drivers of success and profitability of a new surveying business:

(The emphasis on primary drivers) reflects the importance of the right blend of services, clients, working methods, etc, having particular regard to aspects such as costs, profit margins and chargeable time — rather than only fee levels. Indeed, when reviewing the progress of a new surveying business, or considering the growth potential of established businesses, a general contentment with simply being busy can often be found — with attention not being given to factors such as the scope for rejuvenated business development initiatives to secure higher profile clients and greater fee levels/profit margins, and which services and clients are the most profitable.

Business Skills for General Practice Surveyors has not provided a series of business analogies but, as an example, if entering a particular city from the south at around 7.45am, coming off the motorway, two lanes are available. Cars queue in the left-hand lane for 1/4 to 1/2 mile, but the right-hand lane remains clear, except for only a few cars. The lanes do not merge into a single lane, and both lead similarly into the city centre. In the left-hand lane will be people who take the same journey each day and have done so for many years, but have not seen the legitimate opportunity presented by the right-hand lane. In the right-hand lane, are people who have never visited the city before, but know that the opportunity will either be there to drive straight through or, if for any reason it is only possible to get so far, it will be still somehow possible to enter the left-hand lane, and advancement will still have been achieved. Instead of following the herd, they deploy instinct, judgment, vision, focus and strategy, and achieve their aim in approximately 20% of the time of most others — ie 500% as successfully.

It is noticeable that the quality and value of the few cars exploiting the right-hand lane (and therefore perhaps the nature and success in life of their drivers) tends to be far superior to many of the cars in the left-hand lane. Similarly, when there is a long queue in the left-hand lane approaching a roundabout (because most drivers will turn left) and there are one or two, if any, cars in the right-hand lane (turning right or moving ahead), relatively few drivers drive in the right-hand lane, go round the roundabout and then approach the left turn in the same way as would cars coming from other directions. There is no less driving ability in the above situations — the difference lies in the extent of thought, analysis, vision and creativity.

The final word

By way of a final summary, success in the property sector derives from a blend of many factors: technical property knowledge; understanding of business issues/subjects; softer/interpersonal attributes, hard work and some of the finer points considered in this section.

Business Skills for General Practice Surveyors has hopefully provided a grounding in subject areas, and also prompted thought as to how individual surveyors can make the most of their abilities. The surveying profession is an excellent example of where intelligence is not paramount in becoming highly successful, and neither is pure academic/technical knowledge. Instead, analysis and application is key. Work hard, think hard!

Index

For Product Safety Concerns and Information please contact our EU representative GPSR@taylorandfrancis.com Taylor & Francis Verlag GmbH, Kaufingerstraße 24, 80331 München, Germany

Printed and bound by CPI Group (UK) Ltd, Croydon, CR0 4YY

08/05/2025

01864505-0002